NATIONAL ACADEMIES *Sciences Engineering Medicine*

NATIONAL ACADEMIES PRESS
Washington, DC

Greenhouse Gas Emissions Information for Decision Making

A Framework Going Forward

Committee on Development of a Framework for Evaluating Global Greenhouse Gas Emissions Information for Decision Making

Board on Atmospheric Sciences and Climate

Division on Earth and Life Studies

Consensus Study Report

NATIONAL ACADEMIES PRESS 500 Fifth Street, NW Washington, DC 20001

This activity was supported by contracts between the National Academy of Sciences and the Benificus Foundation, Heising-Simons Foundation, and the National Academy of Sciences' Arthur L. Day Fund. Any opinions, findings, conclusions, or recommendations expressed in this publication do not necessarily reflect the views of any organization or agency that provided support for the project.

International Standard Book Number-13: 978-0-309-69114-7
International Standard Book Number-10: 0-309-69114-1
Digital Object Identifier: https://doi.org/10.17226/26641
Library of Congress Control Number: 2022950175

This publication is available from the National Academies Press, 500 Fifth Street, NW, Keck 360, Washington, DC 20001; (800) 624-6242 or (202) 334-3313; http://www.nap.edu.

Copyright 2022 by the National Academy of Sciences. National Academies of Sciences, Engineering, and Medicine and National Academies Press and the graphical logos for each are all trademarks of the National Academy of Sciences. All rights reserved.

Printed in the United States of America.

Suggested citation: National Academies of Sciences, Engineering, and Medicine. 2022. *Greenhouse Gas Emissions Information for Decision Making: A Framework Going Forward.* Washington, DC: The National Academies Press. https://doi.org/10.17226/26641.

The **National Academy of Sciences** was established in 1863 by an Act of Congress, signed by President Lincoln, as a private, nongovernmental institution to advise the nation on issues related to science and technology. Members are elected by their peers for outstanding contributions to research. Dr. Marcia McNutt is president.

The **National Academy of Engineering** was established in 1964 under the charter of the National Academy of Sciences to bring the practices of engineering to advising the nation. Members are elected by their peers for extraordinary contributions to engineering. Dr. John L. Anderson is president.

The **National Academy of Medicine** (formerly the Institute of Medicine) was established in 1970 under the charter of the National Academy of Sciences to advise the nation on medical and health issues. Members are elected by their peers for distinguished contributions to medicine and health. Dr. Victor J. Dzau is president.

The three Academies work together as the **National Academies of Sciences, Engineering, and Medicine** to provide independent, objective analysis and advice to the nation and conduct other activities to solve complex problems and inform public policy decisions. The National Academies also encourage education and research, recognize outstanding contributions to knowledge, and increase public understanding in matters of science, engineering, and medicine.

Learn more about the National Academies of Sciences, Engineering, and Medicine at **www.nationalacademies.org**.

Consensus Study Reports published by the National Academies of Sciences, Engineering, and Medicine document the evidence-based consensus on the study's statement of task by an authoring committee of experts. Reports typically include findings, conclusions, and recommendations based on information gathered by the committee and the committee's deliberations. Each report has been subjected to a rigorous and independent peer-review process and it represents the position of the National Academies on the statement of task.

Proceedings published by the National Academies of Sciences, Engineering, and Medicine chronicle the presentations and discussions at a workshop, symposium, or other event convened by the National Academies. The statements and opinions contained in proceedings are those of the participants and are not endorsed by other participants, the planning committee, or the National Academies.

Rapid Expert Consultations published by the National Academies of Sciences, Engineering, and Medicine are authored by subject-matter experts on narrowly focused topics that can be supported by a body of evidence. The discussions contained in rapid expert consultations are considered those of the authors and do not contain policy recommendations. Rapid expert consultations are reviewed by the institution before release.

For information about other products and activities of the National Academies, please visit www.nationalacademies.org/about/whatwedo.

COMMITTEE ON DEVELOPMENT OF A FRAMEWORK FOR EVALUATING GLOBAL GREENHOUSE GAS EMISSIONS INFORMATION FOR DECISION MAKING[1]

DON WUEBBLES (*Chair*), University of Illinois
KAMAL BAWA (NAS), University of Massachusetts Boston and Ashoka Trust for Research in Ecology and the Environment
GABRIELLE DREYFUS, Institute for Governance & Sustainable Development
ANNMARIE ELDERING, NASA Jet Propulsion Laboratory (Retired)
FIJI GEORGE, Cheniere Energy Inc.
HEATHER GRAVEN, Imperial College London
KEVIN GURNEY, Northern Arizona University
ANGEL HSU, University of North Carolina at Chapel Hill
TOMOHIRO ODA, Universities Space Research Association
IRÈNE XUEREF-REMY, University of Aix-Marseille

National Academies of Sciences, Engineering, and Medicine Staff

RACHEL SILVERN, Program Officer
RITA GASKINS, Administrative Coordinator
ROB GREENWAY, Program Associate
BRIDGET McGOVERN, Associate Program Officer
SABAH RANA, Program Assistant
PATRICIA RAZAFINDRAMBININA, Associate Program Officer
AMANDA STAUDT, Senior Board Director

---

[1] NAS, National Academy of Sciences.
NOTE: See Appendix E for Disclosure of Conflicts of Interest.

BOARD ON ATMOSPHERIC SCIENCES AND CLIMATE[1]

MARY GLACKIN (*Chair*), The Weather Company, an IBM Business (Retired)
CYNTHIA S. ATHERTON, Heising-Simons Foundation
ELIZABETH BARNES, Colorado State University
BRADLEY R. COLMAN, The Climate Corporation
BART E. CROES, California Air Resources Board (Retired)
NEIL DONAHUE, Carnegie Mellon University
ROBERT B. DUNBAR, Stanford University
LESLEY-ANN DUPIGNY-GIROUX, University of Vermont
EFI FOUFOULA-GEORGIOU (NAE), University of California, Irvine
PETER C. FRUMHOFF, Union of Concerned Scientists
ROBERT KOPP, Rutgers, The State University of New Jersey
L. RUBY LEUNG (NAE), Pacific Northwest National Laboratory
ZHANQING LI, University of Maryland
JONATHAN MARTIN, University of Wisconsin–Madison
AMY McGOVERN, University of Oklahoma
LINDA MEARNS, National Center for Atmospheric Research
JONATHAN PATZ, University of Wisconsin–Madison
J. MARSHALL SHEPHERD (NAS/NAE), University of Georgia
DAVID W. TITLEY, U.S. Navy (ret.), The Pennsylvania State University
ARADHNA TRIPATI, University of California, Los Angeles
ELKE WEBER, Princeton University

National Academies of Sciences, Engineering, and Medicine Staff

AMANDA STAUDT, Senior Board Director
APURVA DAVE, Senior Program Officer
APRIL MELVIN, Senior Program Officer
AMANDA PURCELL, Senior Program Officer
STEVEN STICHTER, Senior Program Officer
ALEX REICH, Program Officer
RACHEL SILVERN, Program Officer
MORGAN DISBROW-MONZ, Associate Program Officer
BRIDGET McGOVERN, Associate Program Officer
PATRICIA RAZAFINDRAMBININA, Associate Program Officer
RITA GASKINS, Administrative Coordinator
AMY MITSUMORI, Research Associate
ROB GREENWAY, Program Associate
KYLE ALDRIDGE, Program Assistant
LINDSAY MOLLER, Program Assistant
SABAH RANA, Program Assistant

---

[1] NAE, National Academy of Engineering; NAS, National Academy of Sciences.

Acknowledgments

This Consensus Study Report was reviewed in draft form by individuals chosen for their diverse perspectives and technical expertise. The purpose of this independent review is to provide candid and critical comments that will assist the National Academies of Sciences, Engineering, and Medicine in making each published report as sound as possible and to ensure that it meets the institutional standards for quality, objectivity, evidence, and responsiveness to the study charge. The review comments and draft manuscript remain confidential to protect the integrity of the deliberative process.

We thank the following individuals for their review of this report:

DAVID ALLEN (NAE), The University of Texas at Austin
RONALD COHEN, University of California, Berkeley
INEZ FUNG (NAS), University of California, Berkeley
STEVEN HAMBURG, Environmental Defense Fund
LISA HANLE, Independent Consultant
RODRIGO JIMENEZ, Universidad Nacional de Colombia
AMY LUERS, Microsoft
PAUL PALMER, University of Edinburgh
EMMA STRUBELL, Carnegie Mellon University
FENJUAN WANG, National Institute of Environmental Studies, Japan

Although the reviewers listed above provided many constructive comments and suggestions, they were not asked to endorse the conclusions or recommendations of this report nor did they see the final draft before its release. The review of this report was overseen by **CHRISTOPHER FIELD** (NAS), Stanford University and **SUSAN TRUMBORE** (NAS), Max Planck Institute for Biogeochemistry. They were responsible for making certain that an independent examination of this report was carried out in accordance with the standards of the National Academies and that all review comments were carefully considered. Responsibility for the final content rests entirely with the authoring committee and the National Academies.

Preface

The science is clear—climate change is one of the greatest threats facing humanity. Almost every day the news reports on one or more new extreme weather-related events having serious impacts on some part of the world. Many thousands of observational-based studies have documented the increasing surface, atmospheric, and oceanic temperatures on climate time scales. These observations also show many other aspects of a changing climate—for example, the vast majority of glaciers, including major parts of Greenland and Antarctica, are melting, snow cover is diminishing, sea ice is shrinking, sea levels are rising, oceans are acidifying—and many of these extreme weather events are becoming more intense than they were in the past.

It is also certain now more than ever, based on many lines of evidence, that human activities are largely responsible for these changes in the Earth's climate. Human activities have resulted in emissions that are causing significant changes in atmospheric concentrations of radiatively important gases and particles (collectively referred to as greenhouse gases [GHGs] in this report). As a result of these changes in climate, decision makers are faced with ever-increasing pressure to understand GHG emissions and sinks from human-related and natural sources and their impacts on the Earth's climate as part of emissions mitigation and resiliency strategies. This decision making is important at all spatial scales from local to global if GHG emissions are to be sufficiently reduced to meet agreed upon targets to mitigate the worst effects of climate change.

Improving the characterization of activity-specific contributions to anthropogenic GHG emissions could improve climate predictions, inform decision making at all levels, and enable a more focused and rigorous response to climate change. A more consistent understanding of the impacts of changes in concentrations of GHGs on climate change requires enhanced characterization and quantification of the sources, sinks, and related processes affecting GHG emissions. Enhancing this understanding is therefore fundamental to understanding climate change and the policy responses taken by decision makers.

This fast-track report develops a framework for evaluating anthropogenic GHG emissions information to support decision making. In this study, the Committee examines existing and emerging approaches used in the development and evaluation of global anthropogenic GHG emissions inventories. The developed framework to evaluate emissions information also includes guidance for policy makers about their use in decision making. Given the timely need for gaining an enhanced understanding of human-related emissions, the focus here is on anthropogenic emissions, and this report was completed on an accelerated timeline to be available for international policy discussions later this year.

The National Academies of Sciences, Engineering, and Medicine (National Academies) selected 10 experts in climate processes, human-related emissions, and resulting emissions inventories to be members of the Committee responsible for writing this report. The fast-track process meant that the Committee had a little over 2 months to prepare this report to go to review, and then a few more weeks to respond to the National Academies' external peer-review process. In addressing its tasks, the Committee met twice in person. The Committee also held two information gathering meetings to solicit external input from the international community. The first workshop focused on Greenhouse Gas Emissions Monitoring, Inventories, and Data Integration: Understanding the Landscape with 12 presentations by scientists from the United States, Canada, and Europe. The second workshop, Development of a Framework for Evaluating Global Greenhouse Gas Emissions Information for Decision Making, brought together 29 international experts for "lightning" presentations, and invited moderators led participants through a series of World Café breakout discussions. The Committee also solicited written technical input from the community. These information gathering steps were followed by a series of virtual meetings of the Committee as it worked on the report. Following standard National Academies' procedures, the draft report then underwent a rigorous process of external peer review prior to publication.

I want to finish by thanking the members of the Committee for their extensive efforts and interactions in preparing this report. Every member of the Committee has made seminal contributions to the resulting report. A special thank you goes to those who presented to the Committee and enriched this report in doing so, and to the reviewers who helped us to sharpen and focus the report. Finally, I am sure I speak for my fellow Committee members by expressing our sincere thanks to the entire National Academies' team led by Dr. Rachel Silvern for the huge amount of help they provided to the Committee throughout the process of completing this report.

Don Wuebbles, *Chair*
Committee on Development of a Framework for Evaluating
Global Greenhouse Gas Emissions Information for Decision Making

Contents

SUMMARY   1

1 INTRODUCTION   11
  Quantifying Greenhouse Gas Emissions, 12
  Defining Anthropogenic Greenhouse Gases for This Study, 15
  Defining Greenhouse Gas Emissions Information Scales for This Study, 18
  User and Decision-Maker Needs for Greenhouse Gas Emissions Information
      Across Scales, 19
  Barriers to Widespread Use of Emissions Information, 22
  The Committee's Approach to This Study, 23
  Report Roadmap, 23

2 CURRENT APPROACHES FOR QUANTIFYING ANTHROPOGENIC
  GREENHOUSE GAS EMISSIONS   25
  Activity-Based Approaches, 26
  Atmospheric-Based Approaches, 39
  Hybrid Approaches, 46
  Current Status and Uncertainties of Greenhouse Gas Emissions Estimates, 49

3 STRUCTURAL AND TECHNICAL LIMITATIONS OF THE CURRENT
  GREENHOUSE GAS EMISSIONS INFORMATION LANDSCAPE   53
  Structural Limitations of the United Nations Framework Convention on
      Climate Change National Reporting Process, 53
  Barriers to Capacity Building, 54
  Current Institutions and Their Limitations, 55
  Technical Challenges with Current Approaches, 57

4	**FRAMEWORK FOR EVALUATING GREENHOUSE GAS EMISSIONS INFORMATION**	69
	Pillars in Establishing a Greenhouse Gas Emissions Framework, 69	
	Framework for Evaluating Greenhouse Gas Emissions Information, 73	
	Case Studies Applying the Framework, 79	
	Themes of Framework Evaluation for General Approaches and Case Studies, 92	
5	**RECOMMENDATIONS**	93
	Advancing Greenhouse Gas Emissions Information Capabilities, Trust, and Accessibility, 95	
	Addressing Key Data and Information Gaps and Uncertainties, 97	
	Operationalizing Current Capabilities, 99	
	Striving for Hybrid Approaches, 100	
	Ensuring Usability, Timeliness and Effective Communication of Greenhouse Gas Emissions Information, 101	
	Concluding Thoughts, 103	

REFERENCES 105

APPENDIXES

A	ACRONYMS, INITIALISMS, AND GLOSSARY	133
B	ATMOSPHERIC OBSERVATIONS: METHODS AND EXAMPLES	145
C	CONTRIBUTORS OF INPUT TO THE STUDY	153
D	BIOGRAPHICAL SKETCHES OF COMMITTEE MEMBERS	157
E	DISCLOSURE OF UNAVOIDABLE CONFLICTS OF INTEREST	161

Summary

Current and future changes in the Earth's climate are driven by changes in atmospheric concentrations of radiatively important gases (greenhouse gases [GHGs]) and particles (also called aerosols)—collectively referred to as GHGs in this report. Understanding the emissions of GHGs from human-related (i.e., anthropogenic) and natural sources is critical for national, regional, local, and corporate decision makers seeking to reduce GHG emissions and thwart the most disastrous impacts of climate change. Since 2015, when nations adopted the Paris Agreement to limit global temperature rise this century, governments and the private sector have made commitments to reduce GHG emissions. More than 136 countries, accounting for about 80 percent of total global GHG emissions, have committed to achieving net-zero GHG emissions. GHG emissions information is already being used to plan, track, and assess GHG reduction targets and pledges, and holds the potential to help users make informed mitigation decisions and increase public awareness and accountability.

Three converging trends motivated this report: (1) rapidly increasing demand from a range of users for trusted information about GHG emissions across multiple sectors and geographic scales; (2) development of many new approaches for quantifying GHG emissions that aim to address this increasing demand; and (3) a growing and rapidly evolving institutional landscape, including public, private, and academic entities seeking to provide better GHG emissions information. These converging trends motivated a need for basic criteria or principles that users and decision makers could use when evaluating different types of GHG emissions information.

In this study, the Committee examines existing and emerging approaches used to generate and evaluate anthropogenic GHG emissions information at global to local scales. Ultimately, this report develops a framework for evaluating GHG emissions information to support and provide guidance for policy makers about the use of GHG emissions information in decision making. The Committee's complete Statement of Task is provided in Box S-1.

> **BOX S-1**
> **Statement of Task**
>
> An ad hoc committee of the National Academies of Sciences, Engineering, and Medicine will develop a framework for evaluating global anthropogenic greenhouse gas information to support decision making. Specifically, the committee will
>
> - Describe approaches used to develop global anthropogenic greenhouse gas emissions inventories, including the use of surveys, continuous emissions monitoring systems, monitoring fuel sales, self-reported data, ground- and airborne-based measurements, satellite-based remote sensing data, artificial intelligence and computer vision, proxy and activity data, crowdsourcing, inverse modeling, additional sources of emissions information, and data fusion approaches.
> - Discuss the potential uses and limitations of these approaches, including issues related to
>   - developing global emissions inventories for all anthropogenic greenhouse gas emissions;
>   - the independence, transparency, granularity, internal consistency, comparability, and completeness of data sources;
>   - measures of accuracy and uncertainty of the modeling and analytic tools used to interpret and extrapolate available data;
>   - potential biases introduced by limitations in data availability, model design, or analytic frameworks; and
>   - synthesis, integration, and dynamic updates of the best available information.
> - Provide a framework to evaluate emissions information and inventories, including guidance for policy makers about their use in decision making.
> - Present several case studies to demonstrate how the framework could be applied to evaluate emissions information and inventory approaches and identify strengths and opportunities for improvement for each case study.
> - To the extent possible, identify ways to improve methodological transparency, sustainability and continuity of relevant observations, and product confidence in global anthropogenic greenhouse gas emissions inventories, including key data gaps that could be addressed, improvements needed in models and analytical tools, and opportunities for collaboration among data providers, researchers, regulatory agencies, and decision makers.

Approaches for Quantifying Anthropogenic Greenhouse Gas Emissions

Anthropogenic contributions to the current warming trend globally are driven mainly by the well-mixed GHGs (carbon dioxide [CO_2], methane [CH_4], nitrous oxide [N_2O], fluorinated gases), which account for about 82 percent of present-day warming. About 12 percent of warming comes from non-methane volatile organic compounds and carbon monoxide, which can increase concentrations of atmospheric ozone. About 5 percent of warming is from black carbon. Quantifying emissions of these gases (and particles) is challenging because there are many types of sources and removal processes that have different characteristics, vary over time and space, and encompass countless individual emitters. Some emissions can be measured directly at their source, such as power plants or industrial facilities, whereas other sources are more distributed in space, requiring that emissions be estimated or inferred from other data.

GHG inventories are tools that quantify GHG emissions (or removals) often divided into economic and industrial sectors for a specific place and time. They are developed and used by a range of stakeholders including policy makers, the scientific community, businesses, media, nongovernmental organizations, and the general public. GHG inventories allow policy makers to identify key GHG-emitting sectors and make informed decisions by setting emission baselines, tracking emission changes over time, and assessing emission mitigation efforts. GHG inventories are constructed using a wide range of approaches:

- **Activity-based approaches** (often referred to as "bottom-up" approaches) generally utilize activity data—a term referring to representative indicators or drivers of GHG emissions such as fuel consumption statistics, traffic counts, population, or land area. To achieve an estimate of GHG emissions, these activity measures are transformed using a conversion factor such as an emission factor—the emission or removal of a GHG per unit of activity. For example, the miles of natural gas pipelines would be multiplied by an emission factor representing the methane emissions per mile of natural gas to quantify emissions from a natural gas distribution pipeline system.
- **Atmospheric-based approaches** (often referred to as "top-down" approaches) use atmospheric measurements of GHGs and an understanding of atmospheric transport and chemical processes to infer information on GHG fluxes (emissions and sinks). Surface-, aircraft-, and space-based observations are combined with analysis approaches and models to transform measurements of atmospheric concentrations into estimates of emissions. For GHGs that have both anthropogenic and natural sources, and/or overlapping sources in space and time, attributing estimated emissions to a particular sector or source may require more complex analysis.
- **Hybrid approaches** generate GHG emissions information through the combination and more complete integration of activity- and atmospheric-based approaches, and/or other data sources, data assimilation, or emerging digital technologies (Figure S-1). For example, an activity-based approach using multiple overlapping core datasets could be further constrained by atmospheric-based estimates. Hybrid approaches are nascent and hold the possibility of combining multiple measurement streams and atmospheric- and activity-based approaches to produce more complete and accurate estimates of GHG emissions and sinks.

Data and information products developed using the above approaches have been widely utilized by the scientific and regulatory communities to support emissions reporting, but a number

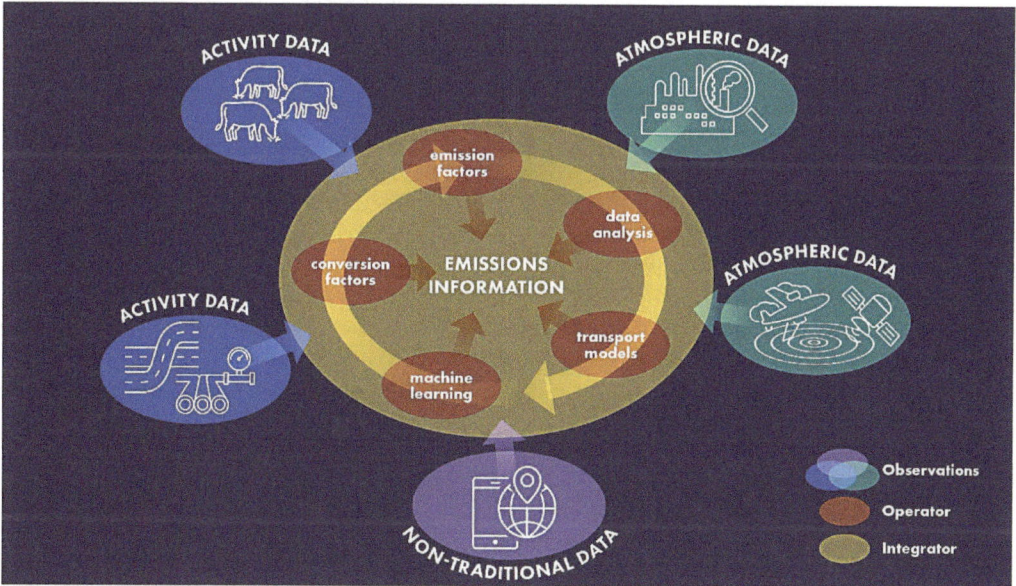

FIGURE S-1 Hybrid approaches generate greenhouse gas emissions information through the combination and more complete integration of activity data (blue), nontraditional data (purple), and atmospheric data (green) that are modified by operator(s) (red) and integrated (gold).

of challenges have limited the usefulness of emissions information to support decision making. For example, activity-based approaches depend on activity data and emission factors that may be out of date or may not be appropriately representative for the system boundary of interest. One of the most widely relied on atmospheric-based approaches, atmospheric inverse modeling, still faces limitations in attributing fluxes to specific sources or sectors to inform mitigation decisions at relevant spatial and temporal scales. While the proliferation of "big environmental data" has considerably advanced the spatiotemporal coverage of atmospheric information, interoperability and transparency are key challenges that can limit the utility of this information, and more research and development are needed to incorporate these data into hybrid approaches.

Framework for Evaluating Greenhouse Gas Emissions Information

As more GHG emissions information becomes available and as more decision makers use this information, a common evaluation framework can help users determine what information products best meet their needs and understand the limitations of that information. A common framework can also provide guidance to researchers for designing more useful and trusted data and information. The Committee has identified six criteria or "pillars" that form a common framework to evaluate current and future GHG emissions information:

1. **usability and timeliness:** information is comparable and responsive to decision-maker needs and available on time scales relevant to decision making;
2. **information transparency:** information is both publicly available and traceable by anyone;
3. **evaluation and validation:** review, assessment, and comparison to independent datasets;
4. **completeness:** comprehensive spatial and temporal coverage of GHG emissions information for the relevant geographic boundary;
5. **inclusivity:** who is involved in GHG emissions information creation and who is covered by the information; and
6. **communication:** methodologies and assumptions are described in understandable forms, well documented, and openly accessible.

In developing this framework and these pillars the Committee seeks to initiate a discussion on the approach and criteria. One of the challenges in developing a framework for assessment is recognizing the many different scales, drivers, and intended uses for GHG emissions information. Figure S-2 summarizes how current capabilities of the three approaches, described above, generally perform relative to the pillars. The evaluation in Figure S-2 focuses on national to global scale GHG emissions information. The qualitative rankings are useful to compare the approaches to each other to identify strengths and opportunities for improvement. Overall, the performance of activity-based approaches spans a wide range but the data tend to be weaker for evaluation and validation. Inclusivity is generally lower, though governmental activity-based inventories tend to perform better for inclusivity and communication. Atmospheric-based and hybrid approaches also tend to rank low for inclusivity and communication, but higher for evaluation and validation. There is room to improve usability and timeliness across all approaches, information transparency for hybrid approaches, and completeness for activity- and atmospheric-based approaches. However, the performance of any given approach will vary depending on the design, implementation, objectives of the effort, and the needs of the information users, as explored in specific case studies in the report.

Drawing on the assessment of activity-based, atmospheric-based, and hybrid approaches against the pillars, the Committee identified needed improvements in current GHG emissions information development capabilities, as described in the following sections. The Committee also

Pillars

Approaches			Usability and Timeliness	Information Transparency	Evaluation and Validation	Completeness	Inclusivity	Communication
	Activity-based	Methods	Medium	High	Medium	Medium	Low	Medium
		Data	Medium	Medium	Low	Medium	Low	Medium
	Atmospheric-based	Methods	Low	High	Medium	Medium	Low	Low
		Data	Medium	Medium	High	Medium	Low	Medium
	Hybrid	Methods & Data	Low	Low	Medium	Medium	Low	Low

FIGURE S-2 Committee's qualitative assessment of the current capabilities of three approaches for quantifying GHG emissions information against the six evaluation pillars. Methods refer to the calculations and mathematical methodologies, and data captures both the inputs (activity data, emissions factors, mixing ratio observations, etc.) and GHG emissions information outputs. Capabilities have been ranked from low (light green) to high (dark green), with medium indicating a high ranking in some instances but a low ranking in others.

recommends striving for hybrid approaches that optimize the integration of individual activity- and atmospheric-based approaches, as a way to provide the best available comprehensive GHG emissions information for users. The Committee's recommendations are intended to apply to the full spectrum of spatial scales relevant to users of GHG information—global, national, regional, local, and facility—and provide clear directions applicable to multiple audiences, empowering different entities to identify specific implementation steps within their means and to meet their needs.

Greenhouse gas emissions information development and evaluation should strive to align with the six pillars: usability and timeliness, information transparency, evaluation and validation, completeness, inclusivity, and communication.

The "pillars" introduced above provide a way to evaluate individual emissions datasets and approaches. Thus, the pillars provide guidance for improving GHG emissions information development and products. These same pillars embody the desired attributes for the institutions that develop GHG emissions information and the broader aspirations for the global, multiscale endeavor of understanding the sources and sinks of GHGs. The application of the six pillars of the framework to both individual datasets and approaches as well as the structures that support the development, provision, and exchange of GHG emissions information would advance the current complex GHG emissions information landscape toward one that could more comprehensively meet the needs of users and decision makers. Strengthening GHG emissions information to satisfy these pillars at local and subnational levels would help to build international coordination and support.

Advancing Greenhouse Gas Emissions Information Capabilities, Trust, and Accessibility

Greenhouse gas emissions information should be better coordinated (e.g., through the creation of a coordinated repository or federation of repositories) across the global community, enabling adherence to a set of minimum common pillar attributes.

A coordinated repository or federation of repositories where GHG emissions information can be hosted, documented, and clearly characterized would be a critical step forward in maximizing use and understanding of GHG emissions data products. A mechanism that brings different types of information together could facilitate the integration of multiple types of data at various spatial scales and make the information accessible to decision makers and users in ways that meet their needs. Such a clearinghouse or federated data center could establish standards and practices, aligned with the pillars, which would enable decision makers and other data users to clearly and quickly grasp individual characteristics and quality of the wide range of GHG emissions information. To maximize adoption and equity, such a coordinated effort should support and integrate information from existing national and international programs to leverage, rather than replace, both established and emerging efforts.

Critical characteristics and functions of a coordinated repository or clearinghouse would operationalize each of the six pillars by including

- Timely information that is transparent and traceable to primary, supporting, and derived datasets;
- Standardization of data formats and metadata to facilitate comparability and interpretability across scales;
- Descriptive documentation of models and estimation procedures in nontechnical language, and in multiple languages, that can be used to enable the use of diverse information, including novel observations and methods from the research community;
- Qualitative (e.g., caveats and limitations) and quantitative (e.g., uncertainties, error characterization) evaluation metrics;
- Databases of key input data and information (e.g., emission factors, activity data, atmospheric observations, models) that would be regularly updated and widely accessible to facilitate information exchange;
- Governance mechanisms that are coordinated, trusted, and designed to be inclusive of the global community and built on the best practices of data governance and information quality;
- Education modules and capacity building for using GHG emissions information and contributing data and estimation results; and
- Mechanisms to support stronger collaborations between GHG, air quality, and meteorological science communities and stakeholders.

Greenhouse gas emissions information providers should clearly communicate underlying data, methods, and associated uncertainties.

While the information clearinghouse or federated repository effort described above would be a longer-term undertaking for the global community, actionable steps can be taken by data providers in the short term to enhance the transparency of GHG emissions information where feasible. Focused resource allocation or government purchases aimed at bringing data and methods into the public domain with standards on transparency and open principles (e.g., FAIR data principles [findable, accessible, interoperable, and reusable]) could have substantial near-term impact on the utility of GHG emissions information. By following many of the same guidelines outlined for a clearinghouse and aligning with the pillars, data providers have the opportunity to facilitate comparability and verification of their data and methods to foster trust between information providers and users.

Addressing Key Data and Information Gaps and Uncertainties

Greenhouse gas emissions information (e.g., observations, data analysis, activity data, emission factors) development at more granular temporal and spatial scales with source-level detail should be accelerated to meet the rapidly increasing needs of cities, states, and provinces for managing their emissions.

Cities, states, provinces, landowners, and the business community, among others, are organizing collectively, enacting mitigation policies, and are in critical need of consistent, standardized, trusted GHG emissions information. Where possible, there are substantial gains associated with extending GHG emissions information development to finer space and time scales, enhancing completeness. Currently, data available on granular spatial and temporal scales are insufficient and there is a need to expand the necessary data resources; this includes activity data, emission factors, and atmospheric observations. Furthermore, enhancing source-level detail—for example by characterizing the entire distribution of emission sources—would strengthen completeness of GHG emissions information and better inform mitigation decisions.

The accuracy and representativeness of all underlying data used to estimate greenhouse gas emissions (e.g., emission factors, activity data) should be further improved.

Many of the data elements, observations, and models used to estimate GHG emissions rely, sometimes by necessity, on large spatial averages or averages that represent well-observed or high-capacity parts of the globe. The Committee recognizes the need to improve both the specific representativeness and resolution across the globe of these key underlying data drivers to strengthen the completeness and accuracy of GHG emissions information. Examples include emission factors that are often fuel averages (i.e., missing true coal quality variation or biomass variation) or tied to countries with well-quantified fuel characteristics; activity data collected for particular countries but used to calculate emissions for other countries even if the data are unrepresentative; and atmospheric monitoring that may only reflect large-scale integration of information or is biased to locations with high scientific capacity.

Operationalizing Current Capabilities

Greenhouse gas emissions estimation research efforts should transition with urgency to operational capabilities with institutions to maintain and ensure longevity.

As the urgency to immediately reduce GHG emissions is increasing, decision makers likewise need the best-available, comprehensive information about emissions as soon as possible. The current timeline and processes to operationalize new data, technologies, or approaches to enable decision-useful strategies is misaligned to meet the above climate goals in a timely manner. Accelerating the transition of research to operations will require scientists, research funders, and data users to identify ways to lower existing barriers to that transition and ways to make new data products more immediately usable. The clearinghouse and other coordination mechanisms recommended above, along with alignment with the pillars, should help make new GHG emissions information usable more quickly.

Some of the approaches for generating GHG emissions information that have been developed within the scientific community are, or are approaching, a level of readiness that would allow for operationalization, resulting in consistent, comprehensive, reliable information that could be better embraced and exploited by decision makers. Much like weather forecasting systems, optimal approaches will combine a more complex suite of observed data to drive optimal estimates of GHG emissions. Iteration between scientists and decision makers is needed to prioritize the continuity and deployment of observing systems and to ensure these approaches are integrated with the decision-making process.

Striving for Hybrid Approaches

Greenhouse gas emissions data collection, modeling, and information development should be designed and implemented to enable a fuller integration and "hybridization" of information and approaches.

Most of the current GHG inventory and information development to date has tended to use single methods or approaches with single-technique observations or data sources. While this was warranted during the development of many of the state-of-the-art estimation techniques, going forward, a "cross-technique" or hybridization of (traditional) approaches and datasets would provide more accurate and comprehensive GHG emissions information by integrating different types of information with more granularity. Some of this work has begun and includes data assimilation and data fusion as well as new machine learning and other nonparametric numerical techniques that leverage new data from private and public satellites, sensors, and other types of activities. Greater synergy between air quality, meteorology, and GHG data collection and analysis efforts would facilitate the development of these hybrid approaches.

Efforts that more fully integrate the traditional activity- and atmospheric-based approaches present a path toward a more complete, complementary approach that overcomes gaps and weaknesses in each approach when used in isolation. Enabling the development of greater integration and hybrid approaches would require designing data collection to fill the most needed gaps. To strive for hybridization is to holistically improve GHG monitoring across scales, approaches, and capacity. Hybrid approaches that mix complementary approaches, datasets, and models and that integrate the needs of end users promise to offer richer, more usable data outputs for decision making.

Ensuring Usability, Timeliness, and Effective Communication of Greenhouse Gas Emissions Information

Greenhouse gas emissions information generators, decision makers, and global stakeholders should engage in an iterative process in a timely manner to ensure the information provided is relevant and useful.

Incorporating decision maker input is critical for information developed to respond to the policy needs of stakeholders and decision makers. The time lag to integrate relevant findings from new research into developing empirical- or measurement-informed inventories limits development and execution of sound mitigation policy, and delays transmittal of appropriate market signals for investments and technology development related to mitigation programs by various stakeholders. Usability and timeliness of GHG emissions information can be enhanced if data producers and users engage in an iterative process, which the clearinghouse or federated repository could support, to facilitate investments in systems that are focused on providing decision support and responsive to an evolving policy-making landscape.

As more users are utilizing and communicating information about GHG emissions, and as more data sources become available, clear expectations about how to evaluate information will help build literacy and shared understanding. Drawing on the pillars in the Committee's framework, Box S-2

> **BOX S-2**
> **Guidance for Evaluating Greenhouse Gas Emissions Information**
>
> The following questions, guided by the pillars, can provide a useful starting point for decision makers, the media, and other stakeholders who are evaluating the credibility and usefulness of new greenhouse gas emissions information:
>
> - Are underlying data, methods, and uncertainties clearly communicated?
> - Over what time period, spatial domain, and for which sources was the information collected?
> - Have the approaches and data been appropriately evaluated and validated?
> - Are multiple data sources or approaches used to support conclusions about greenhouse gas emissions?
> - Have the approaches and resulting data involved locally-based researchers and benefited from stakeholder input and expert review?

outlines key questions that users and communicators of GHG emissions information should pose as they consider how to interpret and use the data.

This report provides a framework for evaluating anthropogenic GHG emissions information that can be adapted as information systems become more complex and to serve a range of decision-making needs. Case studies are provided in the report for several existing approaches to GHG emissions information development, including discussion of advantages and areas for potential improvements. Although it was not the focus of this report, much of the discussion on evaluation and inventories could also be applied to removal processes.

The Committee has made a number of recommendations toward advancing the accuracy and usability of GHG emissions information. By examining existing capabilities and future directions, the hope of the Committee is that this report will help push the global community forward in GHG mitigation decision-making processes.

1

Introduction

Climate change is the most serious environmental challenge currently confronting society. The ongoing COVID-19 pandemic has tested global resolve to address and mitigate threats to the planet and society. At the same time, the pandemic has shown how science and evidence-based policies can help us confront major threats to human well-being. Current and future changes in the Earth's climate depend extensively on changes in atmospheric concentrations of radiatively important gases (greenhouse gases [GHGs]) and particles (also called aerosols). In turn, the concentrations of these gases and particles—collectively referred to as GHGs in this report—are driven by natural and human-related (anthropogenic) emissions and by largely natural removal processes.

Anthropogenic emissions have resulted in large changes in the concentrations of GHGs. Carbon dioxide (CO_2) has the largest impact on current and projected climate and is long-lived in the atmosphere, meaning that human-related emissions of CO_2 cause increases in atmospheric CO_2 concentrations lasting decades to millennia. Nitrous oxide (N_2O) and various fluorinated gases are also important long-lived GHGs. Methane (CH_4) is another major GHG but has a shorter atmospheric lifetime of ~10 years. The methane lifetime is still longer than the time scales of atmospheric mixing so it is relatively well mixed in the troposphere, as are the other long-lived GHGs. As discussed below, anthropogenic emissions of shorter-lived radiatively important gases and aerosols (also called short-lived climate forcers) are also important in determining future climate changes (see Figure 1-1).

GHG emissions information is being used to plan, track, and assess national, subnational, local, and corporate emissions and emission mitigation efforts, often in the context of GHG reduction targets and pledges. In this study, the Committee examines existing and emerging approaches used to generate and evaluate anthropogenic GHG emissions information. Ultimately, this report develops a framework for evaluating GHG emissions information to support and provide guidance for policy makers about the use of GHG emissions information in decision making. The Committee's complete Statement of Task is provided in Box 1-1.

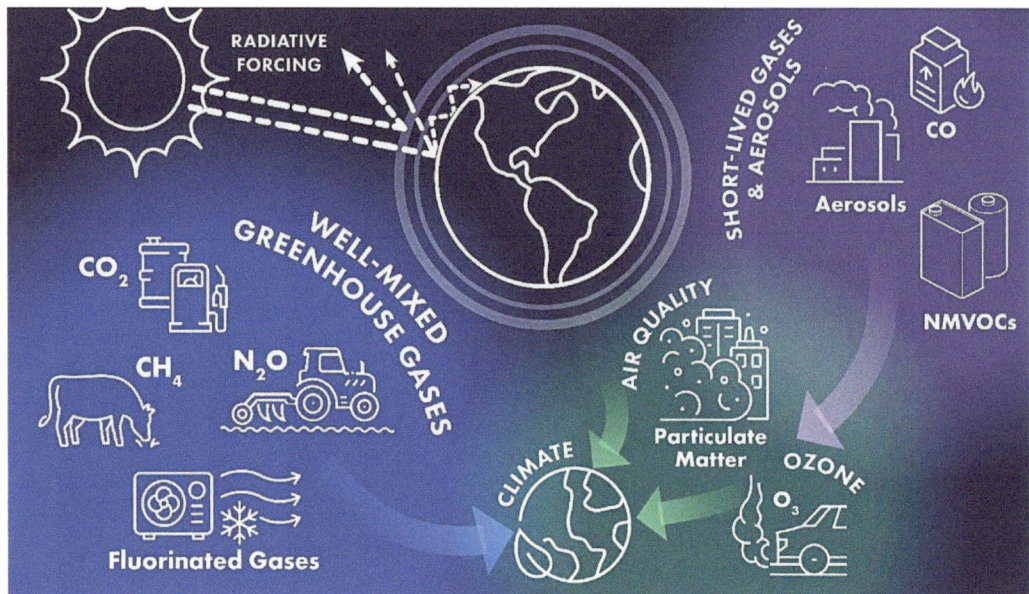

FIGURE 1-1 Well-mixed greenhouse gases (left) and shorter-lived gases and aerosols (top right) impact climate, ozone, and air quality. Greenhouse gases and aerosols impact radiative forcing and changes in global temperature. NOTES: CO_2, carbon dioxide; CH_4, methane; N_2O, nitrous oxide; CO, carbon monoxide; NMVOCs, non-methane volatile organic compounds.

In this report, we will make a distinction between GHG emissions information, GHG inventories, and GHG emissions (see Box 1-2). GHG emissions are the sum of emission or removal values of GHGs. GHG emissions inventories are tools to estimate emissions, often by individual sectors for a specific period of time. GHG emissions inventories can be thought of as a primary subset of GHG emissions information. We define GHG emissions information in this report to include a compendium of emissions-related information content beyond what is often traditionally included in most GHG inventories. GHG emissions information, therefore, might include data used to calculate emissions, an archive of emission factors, or additional attributes of emitters such as ownership, spatial extent, technological characteristics, and other factors such as sociodemographics.

Quantifying Greenhouse Gas Emissions

GHG emissions can be estimated and presented in many ways: by individual GHG, source (e.g., fossil fuel use), geographical region (e.g., city, province, country), economic sector (e.g., transportation, food), or individual emitting infrastructure (e.g., facilities, buildings, roads). Evaluation of GHG emissions information can help determine which policies, technologies, or other interventions could be most effective at reducing emissions. For example, Figure 1-2 shows global GHG emissions (for the well-mixed gases, excluding halocarbons covered by the Montreal Protocol and GHG precursors and aerosols) by gas from 1990 to 2019. It shows that when using 100-year global warming potentials, CO_2 from fossil fuel use and industry contributes about 64 percent of total global emissions and methane emissions account for 18 percent of total global GHG emissions as of 2019, making these sources a priority for reducing emissions. In general, the emissions of CO_2 from fossil fuel burning are well known, whereas there is more uncertainty associated with other sources and gases (as shown by the error bars in the right panel of Figure 1-2).

> **BOX 1-1**
> **Statement of Task**
>
> An ad hoc committee of the National Academies of Sciences, Engineering, and Medicine will develop a framework for evaluating global anthropogenic greenhouse gas information to support decision making. Specifically, the committee will
>
> - Describe approaches used to develop global anthropogenic greenhouse gas emissions inventories, including the use of surveys, continuous emissions monitoring systems, monitoring fuel sales, self-reported data, ground- and airborne-based measurements, satellite-based remote sensing data, artificial intelligence and computer vision, proxy and activity data, crowdsourcing, inverse modeling, additional sources of emissions information, and data fusion approaches.
> - Discuss the potential uses and limitations of these approaches, including issues related to
>   - developing global emissions inventories for all anthropogenic greenhouse gas emissions;
>   - the independence, transparency, granularity, internal consistency, comparability, and completeness of data sources;
>   - measures of accuracy and uncertainty of the modeling and analytic tools used to interpret and extrapolate available data;
>   - potential biases introduced by limitations in data availability, model design, or analytic frameworks; and
>   - synthesis, integration, and dynamic updates of the best available information.
> - Provide a framework to evaluate emissions information and inventories, including guidance for policy makers about their use in decision making.
> - Present several case studies to demonstrate how the framework could be applied to evaluate emissions information and inventory approaches and identify strengths and opportunities for improvement for each case study.
> - To the extent possible, identify ways to improve methodological transparency, sustainability and continuity of relevant observations, and product confidence in global anthropogenic greenhouse gas emissions inventories, including key data gaps that could be addressed, improvements needed in models and analytical tools, and opportunities for collaboration among data providers, researchers, regulatory agencies, and decision makers.

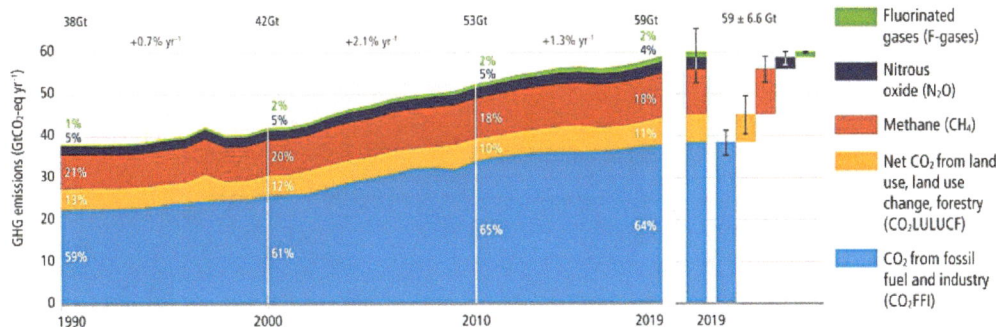

FIGURE 1-2 Global net anthropogenic greenhouse gas (GHG) emissions (for the well-mixed GHGs, excluding halocarbons covered by the Montreal Protocol and GHG precursors and aerosols) from 1990–2019 reported in gigatons CO_2-equivalent per year (GtCO_2-eq yr^{-1}) using 100-year global warming potentials. The right panel shows emissions for 2019 with the associated uncertainties (90% confidence interval) shown as error bars. The fraction of global emissions for each gas is shown in percentages for 1990, 2000, 2010, and 2019. SOURCE: IPCC, 2022b.

> **BOX 1-2**
> **Key Terms Used Throughout the Report**[a]
>
> **Activity-based (or bottom-up) approach:** The activity-based approach (the most common approach used in decision-making applications, to date) involves a wide spectrum of data that indirectly represent emitting processes. In the simplest form, a conversion factor, such as an emission factor, is scaled by the corresponding activity data to estimate emissions. However, this approach can also include direct flux monitoring, ecosystem modeling, and pollution ratio approaches of varying complexity.
>
> **Activity data (AD):** Data on the magnitude of a human activity resulting in emissions or removals taking place during a given time period. Data on energy use, population, equipment count, metal production, land area, traffic data, lime and fertilizer use, and waste arisings are examples of activity data (IPCC, 2019b).
>
> **Air quality:** Generally referred to as air pollution that can have negative effects on human health, plant productivity, or the built environment due to the introduction into the atmosphere of substances (i.e., gases, aerosols) that have a direct or indirect harmful effect (IPCC, 2019b).
>
> **Anthropogenic emissions:** Emissions resulting from human activities of greenhouse gases (GHGs), precursors of GHGs, and aerosols associated with the use of fossil fuels; industrial processes; and agriculture, forestry, and other land use (AFOLU), including deforestation, land use and land use changes (LULUC), livestock production, fertilization, as well as agricultural, industrial, and municipal waste management.
>
> **Atmospheric-based (or top-down) approach:** The atmospheric-based approach estimates emissions using observations of atmospheric concentrations (e.g., ground stations, tall towers, aircraft, and satellites), typically with models that account for atmospheric transport from the emitter to an observation location.
>
> **Atmospheric observations:** In the context of this report, these refer to measurements of the atmospheric concentration or flux of GHGs or tracers.
>
> **Emission factor:** A coefficient that quantifies the emission or removal of a gas per unit activity. Emission factors are often based on a sample of measurement data, averaged to develop a representative rate of emission for a given activity level under a given set of operating conditions.
>
> **Emissions:** Sources and sinks affecting the various GHGs (i.e., their emissions and removal processes).
>
> **Emissions inventory:** A set of estimates of the amount of pollutants emitted into the atmosphere from major mobile, stationary, area-wide, and natural source categories over a specific time period.
>
> **GHG emissions information:** Includes a large collection of information content, including but not limited to emissions inventories. For example, GHG emissions information may include data used to calculate emissions, an archive of emission factors, or additional attributes of emitters.
>
> **Hybrid approach:** The hybrid approach derives GHG emissions information through the combination and more complete integration within and between activity- and atmospheric-based approaches.
>
> ---
>
> [a] A complete list of terms used in this report can be found in the Glossary in Appendix A.

Anthropogenic GHG emissions are receiving increasing attention, particularly at the national, subnational, and corporate levels, driven by the needs of decision makers to effectively plan and achieve emissions reduction commitments. The scientific community has actively studied GHG emissions and atmospheric concentrations to better understand climate forcing and biogeochemical cycles, examining sources and sinks of GHGs that are both anthropogenic and natural. GHG emissions are highly related to the parallel challenges of air pollution, biodiversity loss, and other convergent issues for decision makers and scientists (Box 1-3).

Decision makers need clear, timely, reliable, and complete information about emissions (and removals) that contribute to climate change in order to develop policies that address societal concerns and direct or indirect health and economic impacts. Improving the characterization of anthropogenic GHG emissions could improve predictions of climate change and its impacts, inform decision making at all levels, and enable a more focused and rigorous global response (Gurney and Shepson, 2021). GHG inventories—tools that quantify GHG emission and removal totals or emissions by economic and industrial sectors for a specific place and time—have been constructed for local, corporate, and national-scale decision making. GHG inventories have largely been constructed using activity-based approaches, often referred to as bottom-up analyses (see Box 1-2). Much of this GHG inventory development has been carried out by policy practitioners adhering to standard protocols and approaches. In recent years, the scientific community has developed a wider variety of activity-based approaches and methodologies availing of new data and modeling algorithms (e.g., Bun et al., 2019; Gurney et al., 2009). Examples include the use of "big data" and machine learning, often incorporating very high-resolution imagery and other highly granular techniques (Gurney et al., 2019; Kaack et al., 2022; Rolnick et al., 2022). However, novel approaches from the research community have had little uptake by the decision-making community, due to a number of challenges, described below.

While activity-based analyses are critically important, there have been challenges and limited focus on how information from atmospheric observations can be integrated to further improve the characterization of the GHG emissions and sinks on national, subnational, and local scales to inform specific mitigation strategies. Monitoring and high-resolution models of GHG concentrations in the atmosphere offer the potential to complement and further constrain emissions and budgets independent of activity-based data sources. Most importantly, this atmospheric-based ("top-down") approach to assessing GHG emissions can be integrated with advanced activity-based approaches to potentially provide a more accurate, actionable, and policy-relevant view of GHG emissions relative to changing concentrations of GHGs worldwide.

Defining Anthropogenic Greenhouse Gases for This Study

The most significant anthropogenic contributions to present-day temperature forcing are the major well-mixed gases (CO_2, CH_4, N_2O, fluorinated gases), which account for about 82 percent of present-day warming, or 1.49°C (0.87–2.45°C) out of 1.83°C (0.90–3.12°C), compared to pre-industrial temperatures (IPCC, 2021; see Figure 1-3). Approximately 12 percent of warming is due to emissions of precursor gases and particles: non-methane volatile organic compounds and carbon monoxide (CO), especially because of their effects on the concentration of atmospheric ozone (O_3), and 5 percent due to black carbon emissions. Importantly, about 0.79°C (0.13–1.55°C) of present-day warming is masked by reflective aerosols and aerosol precursors (SO_2, NO_x, organic carbon) and changes in albedo due to land use and irrigation. Definitions of radiatively important gases, particles, and related terms are provided in Box 1-4 and Appendix A.

Future radiative forcing will depend on shorter-lived forcers, including gases and particles, as well as well-mixed greenhouse gases; thus, the definition of GHGs in this report encompasses all

> **BOX 1-3**
> **Converging Information and Challenges**
>
> The development of greenhouse gas (GHG) emissions information is relevant for the scientific community, public, and decision makers to better understand the cascading effects and feedback loops of GHG emissions. Climate, ecosystems, and human society are intimately linked (IPCC, 2022a), impacting each other and creating positive and negative feedback loops (Figure A). For example, human impacts (e.g., deforestation, GHG emissions) can exacerbate climate change while climate change can simultaneously impact human systems. Increasing temperatures push zoonotic epidemics (e.g., COVID-19) and may potentially expand some diseases poleward (Rupasinghe et al., 2022). Climate change may increase extreme weather events, threatening human health and safety (NASEM, 2016) and disproportionately affect the Global South and poor communities in the Global North (Levy and Patz, 2015). GHG emissions are also sources of pollutants that impact human and natural ecosystem health.
>
> Similarly, ecosystems, including biodiversity, can impact climate systems. Many land- and ocean-based systems are becoming sources of GHGs. For example, Amazonian forests, long regarded as important sinks, have been recently shown to be a net emitter of carbon (Gatti et al., 2021). Biodiversity can reduce the negative effects of climate change (Hisano et al., 2018). Restoration, rewilding, and regenerative agriculture are gaining ground as major means to augment biodiversity and enhance ecosystem services, which can act to sequester carbon and other gases. Calls for environmental justice, participatory approaches, and decentralized governance are common to proposals aimed at preventing further degradation of land, oceans, and the atmosphere.
>
> Convergence in goals and approaches is emerging as a major theme in deliberations of international agreements related to climate, ecosystems, and human society. For example, the United Nations' Sustainable Development Goals[a] take a holistic approach, focusing on climate change mitigation as well as ecosystem and human well-being outcomes. The issues discussed in this report benefit from drawing on existing air pollution, weather, and biodiversity information systems and contribute to understanding linkages and impacts on communities and ecosystems. Similarly, the improvement of GHG emissions information may inform and subsequently improve ecosystem and human well-being approaches and outcomes. Furthermore, the availability, usability, reliability, and interoperability of information on ecosystems and other issues are similar to GHG emissions information challenges. Thus, although this report deals with GHG emissions, it has wide implications for policy makers and the scientific community for meeting other global environmental and human welfare challenges.

of these forcers. Emissions of short-lived climate forcers have important implications for mitigation strategies and limiting warming in the near term to below 1.5°C with no or limited overshoot (e.g., Dreyfus et al., 2022; Dvorak et al., 2022). For example, a recent analysis of satellite and other records shows that net aerosol forcing since 2000 has reversed sign from negative to positive and is contributing to accelerating warming (Quaas et al., 2022). Some of these gases and particles have been the focus of air quality emission inventories but have largely not been included in GHG emission inventories. Further consideration of shorter-lived forcers in GHG emissions information may be needed. Additional gases, particles, or indirect GHGs—for example, hydrogen, which can increase concentrations of other GHGs—may also warrant consideration in future GHG inventories if there are sufficient human-related emissions.

A category that may be challenging from the perspective of decision makers is managed vegetation. In general, for mitigation purposes, GHG emissions from human-managed ecosystems have been considered similarly to fossil fuel-related emissions because they are both a consequence of human activities. Managed vegetation refers to planted areas, such as crops, forests, and urban landscapes. Since national and regional mitigation planning is also accounting for these sectors,

INTRODUCTION

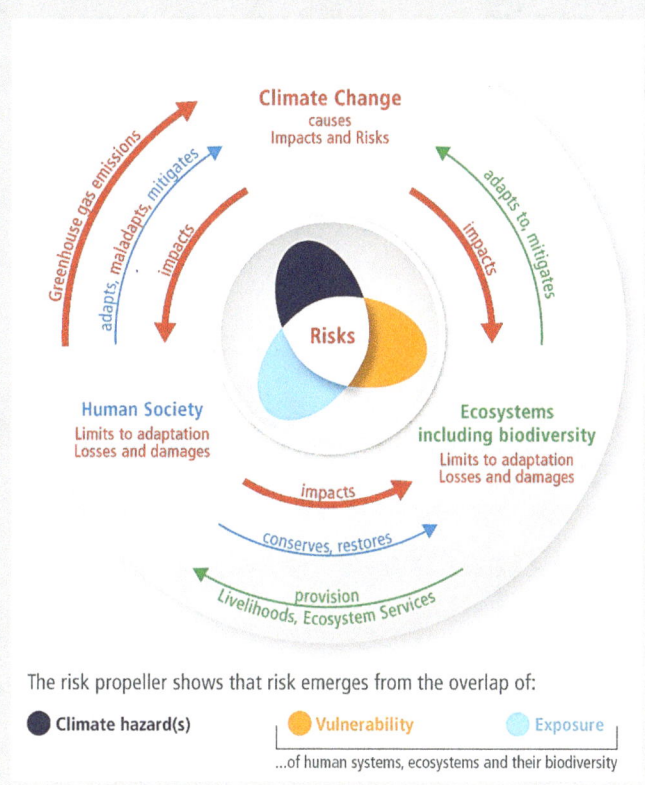

FIGURE A Interactions among the coupled systems: Climate, ecosystems (including their biodiversity), and human society. SOURCE: IPCC, 2022a.

^a https://sdgs.un.org/goals

careful consideration is needed to develop accurate estimates of these human-driven emissions from managed ecosystems. These components are important across a range of scales, from city to regional to national. Improved characterization of managed vegetation would impact estimates of net fluxes, our understanding of the impacts of extreme events, and the overall fidelity of Earth systems models.

While it is well recognized that understanding natural GHG emission sources and sinks (i.e., not managed by humans) in the land biosphere and oceans is essential to quantifying net emissions, the Committee has limited its focus to anthropogenic emissions (defined in Box 1-2).

Many of the methods used to generate GHG inventories can also provide information on natural sources and sinks. The pillars that underpin the assessment framework developed by the Committee are also relevant to information on natural sources and sinks but the primary focus of this report in developing the framework is on the evaluation of GHG information on anthropogenic emissions (and removals). A comprehensive framework with full consideration of approaches specific to natural sources and sinks was beyond the scope of this study.

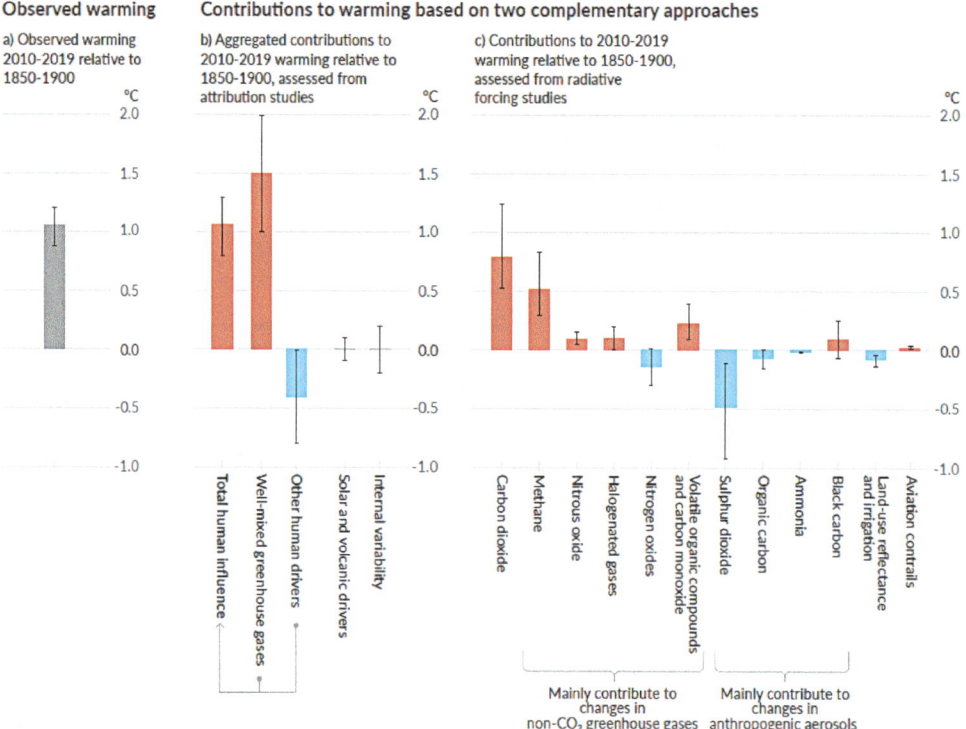

FIGURE 1-3 2010–2019 observed warming relative to 1850–1900 (left) and contributions from attribution studies (middle) and individual radiative forcers (right). SOURCE: IPCC, 2021.

Defining Greenhouse Gas Emissions Information Scales for This Study

The development of GHG emissions information has occurred at many different scales including global, national, regional, urban, and facility spatial scales and annual to hourly temporal scales, discussed below. This report focuses on information at the global scale for which global GHG emissions information and related supporting information are the primary target. However, throughout the report, we include reviews and examples of GHG emissions information at finer spatial scales. The inclusion of this "subnational" GHG emissions information is not exhaustive or comprehensive. Indeed, the sheer volume of GHG emissions information produced at subnational scales has exploded in recent years and attempting to comprehensively capture and reflect on this work would be both challenging and outside the Committee's Statement of Task (Box 1-1).

Nevertheless, we include examples of subnational work throughout the report because many of these efforts offer considerable insight into new inventory techniques, approaches, examples of decision-maker engagement, and lessons learned. Hence, they are included to provide insight and practical examples of particularly successful GHG emissions information elements. This distinction is important because in Chapter 4, an evaluation framework is built to assess current global inventories. The assessment of current capabilities in Chapter 4 is limited to global GHG emissions information and does not reflect the full range of subnational examples and case studies found throughout the report.

> **BOX 1-4**
> **Defining Radiatively Important Gases and Particles and Related Terms**
>
> **Aerosol:** A suspension of airborne solid or liquid particles, with a typical size between a few nanometers and 10 µm that reside in the atmosphere for at least several hours (Allwood et al., 2014). Aerosols may be of either natural or anthropogenic origin and can influence climate through direct and indirect effects.
>
> **Carbon dioxide (CO_2):** The main anthropogenic sources of CO_2 are by-products of burning fossil fuels (such as oil, natural gas, and coal), burning biomass, land use changes, and industrial processes (e.g., cement production). CO_2 is the principal anthropogenic greenhouse gas that affects the Earth's radiative balance (IPCC, 2019b).
>
> **Greenhouse gases (GHGs):** Gaseous constituents of the atmosphere, both natural and anthropogenic, that absorb and emit radiation at specific wavelengths within the spectrum of radiation emitted by the Earth's surface, by the atmosphere itself, and by clouds. This property causes the greenhouse effect. For the purposes of this report, we refer to radiatively important gases and aerosols together as GHGs.
>
> **Halocarbons:** A collective term for the group of partially halogenated species, which includes the chlorofluorocarbons (CFCs), hydrochlorofluorocarbons (HCFCs), hydrofluorocarbons (HFCs), halons, methyl chloride, and methyl bromide. Many halocarbons have large global warming potentials (IPCC, 2019b).
>
> **Methane (CH_4):** The main anthropogenic sources of methane are from three sectors: energy production from fossil fuels, agriculture, and waste (Saunois et al., 2020). Significant anthropogenic emissions within these sectors include emissions from oil and natural gas systems, animal husbandry and paddy rice production, and to a lesser extent landfill waste and wastewater systems.
>
> **Nitrous oxide (N_2O):** The main anthropogenic source of N_2O is agriculture (e.g., fertilizer, soil, and animal manure management), but important contributions also come from wastewater treatment, fossil fuel combustion, and chemical industrial processes (Allwood et al., 2014).
>
> **Ozone (O_3):** The triatomic form of oxygen and a gaseous atmospheric constituent. In the troposphere, ozone is created both naturally and by photochemical reactions involving gases resulting from human activities (e.g., smog). Tropospheric ozone is a GHG (Allwood et al., 2014).
>
> **Particulate matter (PM):** Very small particles emitted during the combustion of biomass and fossil fuels or produced through atmospheric chemical interactions (Allwood et al., 2014). PM may consist of a wide variety of substances.
>
> **Precursors:** Atmospheric compounds that are not GHGs or aerosols, but that have an effect on GHG or aerosol concentrations by taking part in physical or chemical processes regulating their production or destruction rates (IPCC, 2019b).
>
> **Short-lived climate forcers (SLCFs):** Chemically reactive compounds with short (relative to CO_2) atmospheric lifetimes (from hours to about two decades) but characterized by different physiochemical properties and environmental effects.

User and Decision-Maker Needs for Greenhouse Gas Emissions Information Across Scales

Historically, the need for GHG emissions information began at the global and national scales as a necessary ingredient in the international policy-making process. Today, as nations continue to report and use GHG emissions information as part of that process, many have simultaneously used national emissions information domestically, often to inform and support climate-related objectives and targets. In turn, this has stimulated national governments, regional governments, cities, and nongovernmental and private entities to become producers and consumers of GHG emissions information. This section introduces the full range of spatial scales considered in this report—global, national, regional, local, and facility—and the associated context and needs of decision makers and stakeholders for GHG emissions information.

Global and National Scales

Adopted in 1992, the United Nations Framework Convention on Climate Change (UNFCCC) went into force in 1994, establishing the need for GHG emissions information at the global/national scale as part of a formal international treaty (UNFCCC, 1992). GHG emissions information is used to inform GHG policy and mitigation considerations, monitor progress toward commitments, and support the scientific community and public assessment of GHG emissions and removals (ClimaSouth, 2014; Lindroth and Tranvik, 2021; Yona et al., 2020). The UNFCCC includes a formal process whereby UNFCCC Parties develop and update GHG inventories based on guidelines produced by the Intergovernmental Panel on Climate Change (IPCC). Tracking GHG inventories over time and continually assessing progress toward agreed upon mitigation targets was of primary importance and emphasized regular, continual GHG budget quantification. GHG emissions information has also been used within the international negotiating process itself with countries discussing targets and mitigation options contextually grounded with ongoing numerical assessment of national emissions and their sector composition.

Outside of the UNFCCC reporting framework, GHG emissions information development at the global and national scale has a long history dating back to the 1970s (Keeling, 1973; Marland and Rotty, 1984). While the development began for primarily scientific research reasons (e.g., understanding the global carbon cycle, climate change projections), the outcome of the early GHG inventory efforts has provided methodological foundations to the emission calculations and approaches suggested in IPCC guidelines and is increasingly needed as both alternative and complementary information to the UNFCCC process. At a national level, decision makers need national quantification of GHG emissions and removals to understand their national GHG landscape, plan mitigation targets and actions, and place their national budgets within the global context to make measurable progress on limiting climate change.[1] In addition, various nongovernmental organizations, the media, think tanks, the business community, and others have long used and archived national GHG emissions information to track progress and maximize emissions transparency (Yona et al., 2020).

Subnational Government Scales

Since the 2015 Paris Agreement formally recognized the contributions of "all levels of government and various actors" to achieve global climate mitigation goals, the number of cities, state and regional governments, and private actors pledging and implementing their climate actions has grown (Hale, 2020; Hsu et al., 2018; IPCC, 2022c). A need for GHG emissions information at these smaller subnational scales also originated mostly independent of the international policy process as subnational entities have taken on mitigation planning and bi/multilateral policy arrangements (Rosenzweig et al., 2010). The UNFCCC's Global Climate Action Portal recorded 11,355 city and 270 regional governments pledging some form of individual climate mitigation pledge or participation in a transnational climate network, such as the Global Covenant of Mayors for Climate and Energy, which is the largest city-actor initiative with nearly 12,000 cities participating as of July 2022.

Urban-scale (i.e., neighborhoods, communities, metropolitan areas) emissions occupy a unique niche among the subnational decision makers and stakeholders. First, cities account for almost three-quarters of energy and transportation-related atmospheric CO_2 emissions and that share is projected to grow substantially in the coming decades (Gurney et al., 2022; Seto et al., 2014). This reality in combination with the acknowledgment that low-emission development is often consistent with a variety of other co-benefits (e.g., air quality improvement) has pushed cities to independently pursue GHG emissions mitigation (Hsu et al., 2015; Rosenzweig et al., 2010; Watts, 2017).

---

[1] https://unfccc.int/process-and-meetings/the-paris-agreement/the-paris-agreement

Within these networks and cooperative climate initiatives, subnational governments are frequently required to regularly report emissions inventories detailing progress toward climate mitigation goals. Subnational characterization of GHG emissions also finds use among a larger swath and complex array of the decision-making world. For example, citizens, local environmental organizations, and local media are active in using GHG emissions information in planning and strategizing climate action activity (Hsu et al., 2020).

Despite these efforts, however, existing subnational estimates of GHG emissions can be incomplete, contain potential flaws, be difficult to intercompare, and are primarily skewed toward cities in the Global North (i.e., the United States, Canada, United Kingdom, European Union, Singapore, Japan, South Korea, Australia, and New Zealand) (Gurney et al., 2021; Ibrahim et al., 2012; Wei et al., 2021). One challenge of estimating emissions at disaggregated spatial scales is that many of the methods were developed to calculate GHG emissions and removals at the national level. Subnational decision making needs subnational GHG emissions information, but at these scales, there are reporting challenges associated with system boundaries (e.g., emissions leakage, lateral transfers, scope definitions) and multigovernance arrangements, among others (Gurney et al., 2015; Hutyra et al., 2014). While there have been some efforts to standardize subnational emissions reporting and align guidance with national inventory methodologies such as the IPCC's 2019 refinement of the 2006 national inventory guidelines (e.g., Fong et al., 2014), subnational governments have discretion in the ways that they both estimate and report GHG emissions.

Facility and Corporate Scales

The facility or individual business scale has also emerged as an important consumer of GHG emissions information, often because of corporate energy transition strategies and project financing related to climate impacts in alignment with the Paris Agreement. Over 40 percent of the world's 2,000 largest publicly traded companies have committed to net-zero or other climate targets.[2] Over 90 percent of the S&P 500 and 80 percent of the Russell 1000 companies issue corporate environmental, social, and governance reports with GHG and carbon footprints accounting for emissions reductions as key performance indicators.[3] The success of these commitments relies on accurate GHG inventories and transparent reporting of their progress. Contrasting governments' geographic focus, this scale places particular emphasis on supply-chain approaches and ownership boundaries. For example, the Task Force for Climate-related Financial Disclosures provides a guide for firms to properly disclose their climate-related risks and opportunities and calls for businesses and financial services companies to regularly evaluate and update their analyses of these risks and opportunities (e.g., UNEP, 2020). Corporate or facility-level accounting forms the underlying basis to conduct such analysis.

An inaccurate baseline or even prospective targets based on poor, inaccurate, or incomplete accounting could result in a lack of trust and reputational damage, including claims of "greenwashing" or ineffective deployment of capital and resources (Pfadt-Trilling and Fortier, 2021). An analysis of public corporate reports indicates shortcomings of incomplete inventories (Day et al., 2022; Tollefson, 2022). Hence, establishing corporate commitments or assessing progress based on inaccurate or incomplete information may limit the ability of a reasonable investor and other stakeholders to make informed decisions on the GHG performance of the company (Cannon et al., 2020).

---

[2] https://zerotracker.net/#companies-table
[3] https://www.yahoo.com/now/92-p-500-companies-70-140530175.html

Barriers to Widespread Use of Emissions Information

Decision makers require timely and accurate information by which to base management decisions, as well as to evaluate the effectiveness of policy responses. However, information uptake and translation of GHG emissions data and information to actionable knowledge and insight can fail to occur for a variety of reasons:

- **Capacity:** Decision makers must have the capacity to either collect and analyze their own emissions data or utilize datasets generated from other sources to inform action. The capacity for emissions information data collection may be limited due to a lack of resources, technical know-how, or a range of other reasons (e.g., Andres et al., 1996, 2012, 2014). In terms of the ability to utilize others' data and information, analytical capacity, defined as the ability of decision makers to analyze information, and the application of research methods and modeling techniques (Howlett, 2009), may be a limiting factor. In the age of big data and information, decision makers can frequently become overwhelmed with the vast amount of data and information available. Without considering the analytical capacity of decision makers or end users of information, "overloading" capacity can lead not only to the failure of data and information uptake but also risks mismanagement of the original problem (Dietz et al., 2003). Decision makers may also lack relevant expertise or background to know how to analyze or utilize data or information when presented.
- **Transparency:** The processes and methods used in developing GHG emissions information give credibility to the dataset and are important for understanding potential uncertainties. This underlying information may not be provided to decision makers in a transparent manner or not clearly communicated and, thus, decision makers may not be comfortable using a given emissions dataset.
- **Accessibility:** Data may not be available in formats that are easily accessible or usable for decision makers. Documents may be hard to extract and use or can be unwieldy in size or format, making them difficult to download, format, and analyze, depending on the capacity of users. This challenge is more general to Earth science data (Carlson and Oda, 2018). The lack of a common framework for reporting and analysis contributes to this barrier because it privileges experts and prevents decision makers from learning from each other.
- **Trust:** Users need to find the GHG emissions information trustworthy. With competing estimates and lots of political and financial interests involved in emissions analyses, there can be a lack of trust or at least suspicion that can create missed opportunities.
- **Timeliness:** GHG emissions inventories developed according to the IPCC's Tier 1 or 2 methodologies (IPCC, 2006) are primarily based on activity data (e.g., fuel consumption) that take time and resources to collect and aggregate (Andres et al., 2012). As a result, emissions inventories are generally only available with a few-year time lag and may not be indicative for decision makers requiring real-time information on which to base policy decisions, or individual entities developing their respective mitigation actions. For example, the latest national-level information for developed countries is the 2022 submission to the UNFCCC (UNFCCC, 2022a), containing data for 2020; for developing countries, the latest year of data availability varies but is never more recent than 2–3 years before the year of submission (UNFCCC, n.d.-b, c). On the other hand, near-real-time emission estimates can be more timely but less accurate (Oda et al., 2021).
- **Relevance:** Data may not be, or not be perceived to be, relevant for a decision maker, limiting its application and uptake. There may be political ideologies that hinder decision makers from utilizing data because policy makers may interpret data based on past attitudes or beliefs that may deem certain information irrelevant (Arnautu and Dagenais,

INTRODUCTION

2021). In other instances, data may not be available at a temporal or spatial scale relevant to a decision maker. For example, a national emissions inventory may not be relevant for a local government or private business to base decisions on; therefore, the data are not utilized. Projections of emissions and comparisons against pledges are often insufficiently granular to assess the relevance of the transition mitigation strategies of individual stakeholders. For example, the United Nations' Emissions Gap Reports[4] are useful in modeling scenarios and identifying gaps in pledges, but they lack granularity (e.g., by not mapping the pledges against specific sectors). Moreover, use of CO_2 equivalents is an inaccurate representation of resulting climate effects and leads to ambiguity when assessing which GHGs are contributing the greatest temperature impacts. As such, annual GHG inventories reported under UNFCCC guidelines are encouraged or required to report the contribution of each gas by mass (IPCC, 2006). The reporting of GHGs separately allows GHG inventories to be used to understand the implications and efficacy of policies over different periods of time.

- **Awareness:** Decision makers may not necessarily be aware of the data available to them to base policy decisions on. Since data collection and analysis technologies are constantly evolving, it may be challenging for a decision maker to constantly stay up to date on the latest data. The quality of the data and how it compares with other data from an agency or institution may also affect the adoption or uptake of a dataset. Limited communication between decision makers and data providers or scientists is another obstacle to awareness, trust, and transparency.
- **Long-term support:** Just as long-term monitoring is important to determine trends, long-term funding support for GHG emissions measurements and analyses is also important for informing decision making. The challenge is to maintain support for measurements and analyses that extend beyond political cycles.

The Committee's Approach to This Study

In addressing its tasks, the Committee met twice in person in Washington, DC, in June 2022. The Committee held two information-gathering meetings to solicit external input from the international community. The first meeting was held on June 2, 2022, hosted by the National Academies' Board on Atmospheric Sciences and Climate, on Greenhouse Gas Emissions Monitoring, Inventories, and Data Integration: Understanding the Landscape with 12 presentations by scientists from the United States, Canada, and Europe. On June 27–28, 2022, the Committee held a workshop on Development of a Framework for Evaluating Global Greenhouse Gas Emissions Information for Decision Making. This workshop had 29 international experts give "lightning" presentations and invited moderators led participants through a series of World Café breakout discussions. The Committee also solicited written technical input from the community. These information-gathering sessions were followed by five virtual meetings in July and August 2022 during which the Committee deliberated and wrote the report. Following standard National Academies' procedures, the draft report then underwent a rigorous process of external peer review before publication.

Report Roadmap

Chapter 2 describes approaches for the development of GHG inventories that are used to set emissions baselines and measure emissions changes. In this report, the Committee considers three approaches: (1) activity-based approaches that utilize activity data as representative indicators of

---
[4] https://www.unep.org/resources/emissions-gap-report

GHG emissions; (2) atmospheric-based approaches that use measurements of atmospheric concentrations to infer information on emissions and sinks; and (3) hybrid approaches that functionally integrate multiple data sources and approaches.

Chapter 3 describes the institutional and technical limitations of current approaches that have inhibited their usefulness in decision making.

Chapter 4 develops the framework for evaluating GHG emissions information. The framework includes six "pillars" that serve as criteria to qualitatively assess the extent to which a given information system or inventory performs. Chapter 4 also describes how current approaches perform relative to the pillars and provides case studies exemplifying how the framework could be utilized to evaluate and improve current and future efforts.

Chapter 5 makes actionable recommendations for the short and long term. These recommendations aim to enhance the capabilities of GHG emissions information to fulfill the six pillars of the framework to provide trusted, usable, and policy-relevant information for a wide range of users and decision makers.

2

Current Approaches for Quantifying Anthropogenic Greenhouse Gas Emissions

Greenhouse gas (GHG) inventories are tools that quantify GHG emissions (or removals) often divided into economic and industrial sectors over a specific time period. They are most often resolved to the nation, province, or city scale but can also be resolved down to individual emitting infrastructure (e.g., facilities, buildings, roads), or aggregated over multiple facilities across geographies or jurisdictional boundaries. Many global inventories are spatially resolved to regular "gridded" form, a common practice at the global scale, allowing for integration with atmospheric modeling systems. GHG inventories typically span multiple years, with annual resolution, but many will resolve to finer temporal scales such as month, day, or hour. GHG inventories are developed and used by a range of stakeholders including policy makers (at multiple scales), the scientific community, businesses, media, nongovernmental organizations (NGOs), and the general public. By quantifying GHG sources and sinks, GHG inventories can inform emission baselines or spatial distributions, track emission changes over time, and assess emission mitigation efforts for a specific entity over a period of time (Schaltegger and Csutora, 2012). Inventories can also form the basis of projections and future emission scenarios, further assisting with mitigation planning and tracking.

GHG inventories are constructed using a wide variety of approaches. Broadly, in this report, we classify them into three groups: (1) activity-based approaches, (2) atmospheric-based approaches, and (3) mixed or "hybrid" approaches. The Committee chooses not to use the more familiar "bottom-up" and "top-down" terminology. Activity-based approaches are most familiar to decision makers and have been most commonly deployed for decision making purposes. Atmospheric-based approaches have seen extensive development in the research community. While atmospheric-based approaches have had less application to decision-making, to date, increasingly their application to decision making is being discussed, often as a complement to activity-based approaches. Finally, mixed or hybrid approaches have seen only recent development, but offer a promising means to build more relevant and accurate inventories in the future. We describe each of these approaches below, noting source data, general methods, scope, and current examples.

Activity-Based Approaches

Activity-based approaches (often referred to as "bottom-up" approaches) encapsulate a wide variety of data sources and methods. The common element among the many activity-based GHG inventories is the general use of activity data, a term referring to representative indicators or drivers of GHG emissions within system (or model) boundaries, such as fuel consumption statistics, traffic counts, local air pollution quantities, population, or vegetative cover. The type of activity data employed in GHG inventory estimation efforts is often unique to the economic sector, technology, and the specific GHG. In this report, we also include direct flux monitoring such as a continuous emissions monitor (typically placed in an effluent stream), as part of activity-based approaches. Even though continuous emissions monitors measure gas concentrations, these are not considered traditional atmospheric measurements and require additional information (mass flowrate) to be used to estimate an emissions amount. Hence, they are grouped with activity-based (or other "bottom-up") approaches in this report. By contrast, we consider eddy flux measurements, which incorporate atmospheric dynamics, part of atmospheric-based approaches.

To achieve an estimate of a GHG flux, these activity measures must be transformed using an "operator" or emission factor (a coefficient that represents the emission or removal of a GHG per unit of activity) (Figure 2-1).

This general approach is often represented *in its simplest form* as

$$emissions = AD \times EF \qquad (1)$$

where *AD* represents an activity measure and *EF* an emission factor. This simple representation (Equation 1) assumes that emission factors are constant and representative for a given activity for the duration of consideration. Specifically, activity-based approaches may fail to describe complex

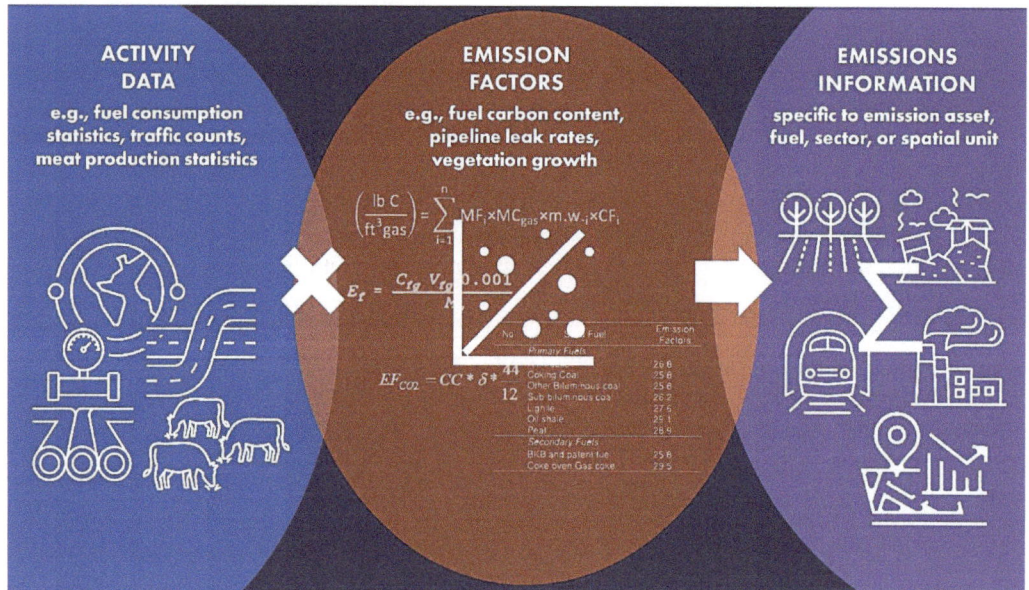

FIGURE 2-1 Activity-based approaches multiply activity data (representative indicators or drivers of greenhouse gas [GHG] emissions, such as fuel consumption statistics, population, equipment count, traffic counts) by an emission factor (a coefficient that represents the emission or removal of a GHG per unit of activity) to produce GHG emission totals or emissions by sector.

processes such as emissions and removals associated with managed vegetation, unintended emissions or disruptive emissions (e.g., equipment failure, natural gas pipeline leaks), and wartime military activities that require more complex modeling constructs. Additionally, some emission factors are well characterized with measures of uncertainty (e.g., the carbon content of natural gas has been measured widely and its regional variation well understood), while others are either established from small samples or specific application settings and are poorly characterized or represent mean values (e.g., N_2O from soils). Because activity-based approaches are most often sector, fuel, and technology specific, they comprise a mixture of both simple and more complex versions of Equation 1, and can incorporate direct flux monitoring.

Production (Direct) Versus Consumption (Indirect) Perspectives

GHG inventories can be expressed as "production" (i.e., direct) or "consumption" (i.e., indirect) inventories, depending on how the system boundary is defined for the GHG accounting problem (Peters, 2008). Production inventories tie emissions to the geographic location where they enter the atmosphere (Figure 2-2). These emissions are often also referred to as "Scope 1" emissions, a term that emanates from the language originating with corporate and organizational GHG accounting (WRI/WBCSD, 2004). The consumption-based GHG perspective refers to emissions that occur as a result of consumptive activity, much of which may occur in geographically distant locations outside of an entity's territorial boundary. Therefore, consumption emissions incorporate emissions associated with the supply chain for all goods and services (often referred to as "Scope 3" emissions; Figure 2-2) and may employ, for example, life-cycle accounting techniques to quantify emissions (Chen et al., 2019). Although this approach is often thought of as incorporating "upstream"

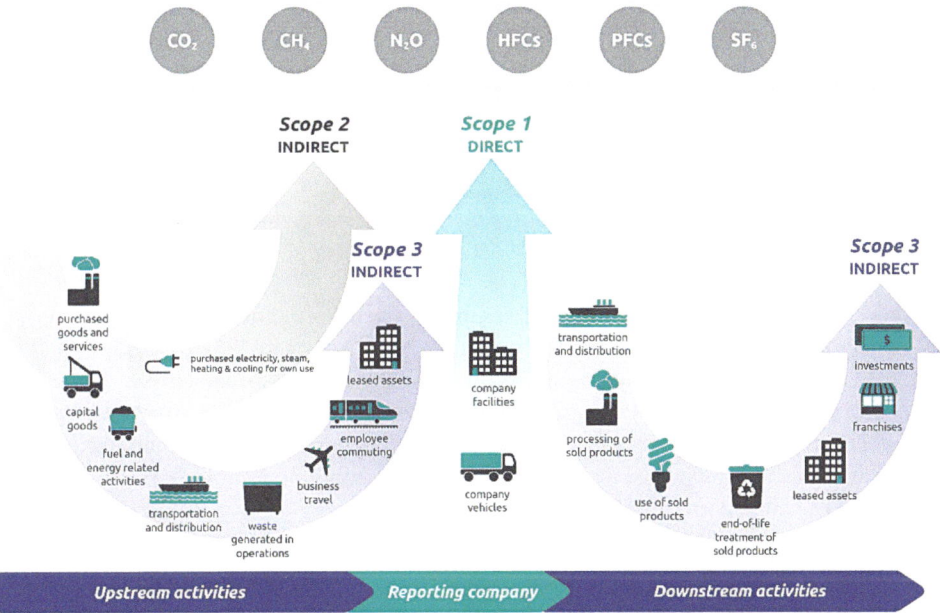

FIGURE 2-2 Classification of Scope 1, 2, and 3 emissions in production and consumption greenhouse gas emission inventories. NOTES: CO_2, carbon dioxide; CH_4, methane; N_2O, nitrous oxide; HFC, hydrofluorocarbon; PFC, perfluorocarbon; SF_6, sulfur hexafluoride. SOURCE: WRI/WBCSD, 2011.

emissions, it also reflects "downstream" activities such as waste production, transport to landfill or combustion locations, and end use of a product. Scope 2 emissions are specifically related to the assignment of GHG emissions from electricity, heat, or steam production from a power plant to the point of consumption (e.g., household, office building, factory; Figure 2-2).

The distinction between Scope 1 versus Scope 2 and 3 accounting perspectives tends to be a more critical consideration as the spatial scales decrease. The accounting perspective is also critically important when considering approaches to quantify GHGs. For example, because atmospheric-based approaches only "see" emissions from the Scope 1 or production perspective, incorporating atmospheric-based approaches with consumption-based GHG accounting can be challenging to integrate.

Survey of Current Global/National Activity-Based GHG Inventories

Many activity-based GHG inventories have been developed over the past four decades. We first distinguish a hierarchy for describing these GHG inventories. We consider four size classes: (1) global/national, (2) national/regional, (3) local/urban, and (4) facility (see also Chapter 1) and their activity-based approaches below. The Statement of Task for this report focuses on the development of an evaluation framework for GHG information at the global/national scale, but here we provide a more complete general description across the scales currently considered in GHG emissions information efforts to provide context and useful case study examples.

UNFCCC GHG Inventories

The United Nations Framework Convention on Climate Change (UNFCCC) established the need for GHG emissions information at the global/national scale by requiring national governments to regularly report inventories. Participating Parties have employed the Intergovernmental Panel on Climate Change (IPCC) 1996 and 2006 guidelines in the preparation and submission of these inventories (IPCC, 2006; UNFCCC, 2014). Recognizing the limited institutional capacity of various countries in addressing methodological complexities, the IPCC guidelines specify a "tiered" GHG estimation approach (i.e., Tier 1, 2, 3) in which the complexity and presumed accuracy increase from Tier 1 to 3. The aim is to create inventories that are "Transparent, Accurate, Complete, Consistent, Comparable, and efficient in resource use" (IPCC, 2006). Tier 1 defaults generally involve the use of country-average data, while Tiers 2 and 3 involve more refined or granular estimation methods. In general, the IPCC emissions inventory estimation methods involve the use of activity data and corresponding emissions per unit of activity (Rypdal et al., 2006). The UNFCCC reporting process is summarized in Figure 2-3.

Under the Paris Agreement, all signatories are required to employ the Enhanced Transparency Framework (ETF)[1] by 2024 to report, in a uniform basis, national GHG inventories and support collective evaluation efforts, including the every 5-year Global Stocktake (Box 2-1; see also Case Study 4.1). The ETF incorporates capacity considerations of developing countries while aiming to avoid undue burdens on the Parties (UNFCCC, 2020).[2] The Capacity-building Initiative for Transparency provides support to developing countries' commitments under the ETF, including the development of national inventories. Under the ETF process, all inventories will undergo review by

---

[1] The ETF was established by Article 13 of the Paris Agreement.

[2] Developed countries are required to report seven gases (CO_2, CH_4, N_2O, HFCs, PFCs, SF_6, and NF_3). Developing countries that need flexibility due to their capacity level are required at a minimum to report emissions of CO_2, CH_4, and N_2O, and any of the "F gases" (HFC, PFC, SF_6, and NF_3) if included in the Party's nationally determined contributions (NDCs) or if they have F gases under Article 6 or previously reported them (UNFCCC, 2019, para. 48). Parties should report international aviation and marine fuel bunker emissions and should not include those emissions in national totals.

CURRENT APPROACHES FOR QUANTIFYING ANTHROPOGENIC GREENHOUSE GAS EMISSIONS

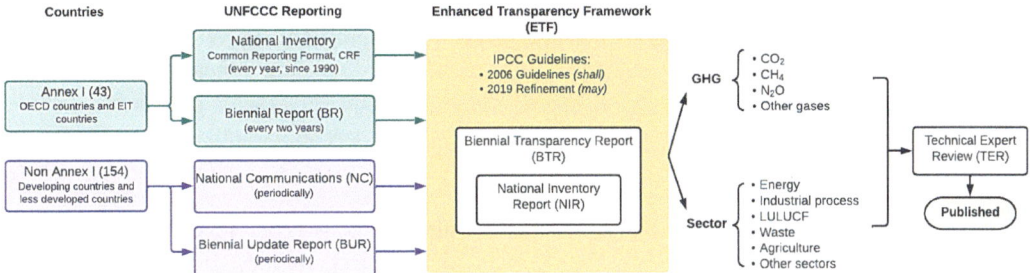

FIGURE 2-3 Structure of national reporting processes for Annex I (green) and non-Annex I (purple) countries and processes for future submission under the Enhanced Transparency Framework (gold shading). NOTE: LULUCF, land use, land-use change and forestry. SOURCE: Adapted from Deng et al., 2022.

BOX 2-1
Global Stocktake of the Paris Agreement

The Global Stocktake (GST) of the Paris Agreement is a once every 5-year process to collectively assess progress toward achieving the climate change goals established in the agreement itself (Article 14, Paris Agreement; UNFCCC, 2015). The first GST is slated from 2021 to 2023 and will repeat every 5 years, aiming to evaluate progress on three major elements of the Paris Agreement and United Nations Framework Convention on Climate Change (UNFCCC): mitigation, adaptation, and means of implementation and support, including financing commitments that provide support to developing countries to implement their nationally determined contributions (NDCs) (UNFCCC, n.d.-a). According to the UNFCCC, the GST "considers the social and economic consequences of response measures and efforts to address loss and damage," which pertains to climate change impacts disproportionately experienced by developing countries as the result of greenhouse gas (GHG) emissions primarily the historic responsibility of developed, industrialized countries (UNFCCC, n.d.-a). The GST further emphasizes equity and "makes use of the best available science," intending to increase the overall ambition of climate action to close the gap between pledged NDCs and emissions levels consistent with 1.5° temperature pathways (UNFCCC, n.d.-a).

The UNFCCC relies on national inventories by both developed and developing countries to support the GST. The UNFCCC report released in March 2022 details:

- GHG emissions by sources and removals by sinks;
- Total aggregate GHG emissions, excluding land use, land-use change, and forestry, with sectoral and specific gas breakdowns, between 1990 and 2019, for developed countries and 55 developing countries; and
- Adequacy of policy efforts and national ambition to achieve global climate goals.

At the 2021 Conference of the Parties (COP)-26 Glasgow Climate Summit, the UNFCCC Secretariat released an NDC Synthesis report that determined that enhanced ambition mitigation efforts countries pledged in the lead-up to the Summit would amount to 2.7° warming by the century's end.

a group of technical experts to ensure reporting is consistent with international rules. Parties will track progress toward their country's nationally determined contributions (NDCs)—non-binding national plans that include climate mitigation, adaptation, and financing commitments (including climate-related targets for GHG emission reductions)—before the publication of reports.

Global/national GHG Inventories

Beyond UNFCCC inventory reporting, many national/global GHG inventories have been developed with varying characteristics. For example, all the energy-related carbon dioxide (CO_2) emissions in various datasets rely on energy statistical data from a small number of energy-reporting sources (Figure 2-4). The major four energy data archiving agencies (Energy Information

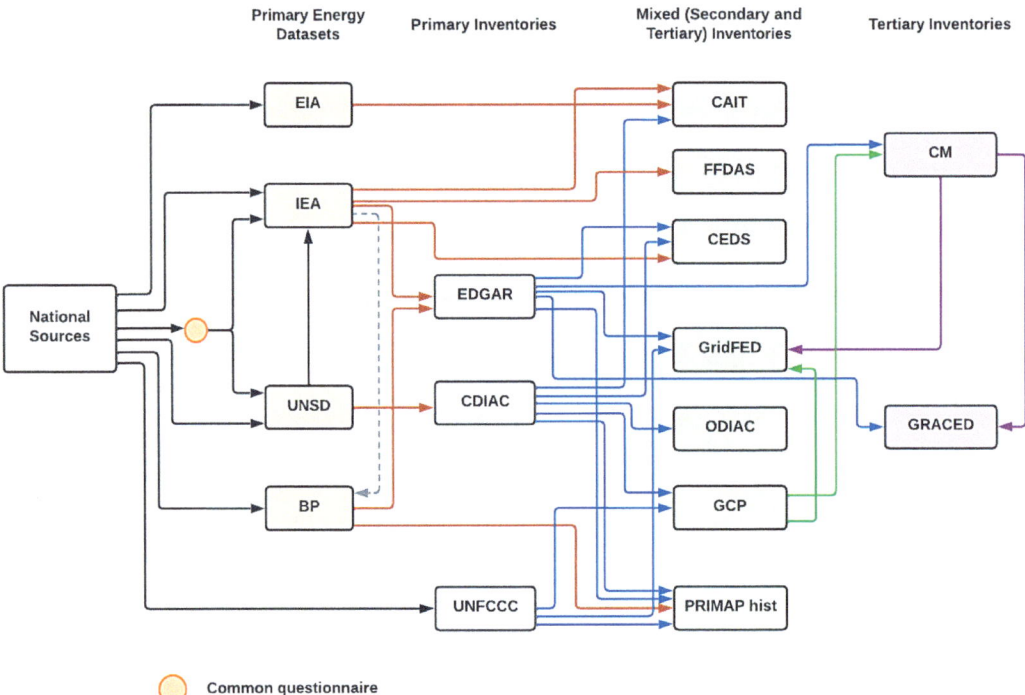

FIGURE 2-4 Example of the sources and the flow of fuel statistics information from national sources through selected global greenhouse gas (GHG) datasets for energy-related emissions. The colored arrows simply trace one column to the next. "Primary energy datasets" (yellow) refer to energy datasets collected directly from national sources (EIA, IEA, and BP also calculate GHG emissions). "Primary inventories" (blue) may combine two or more of the energy datasets. "Mixed secondary/tertiary inventories" (green) refer to datasets that use a mixture of primary and secondary data sources. "Tertiary inventories" (purple) refer to datasets that use sources from primary inventories only and typically provide additional space/time resolution or specific data additions (not shown). BP (formerly British Petroleum) uses non-fuel shares from the International Energy Agency (IEA) (dashed line). NOTES: EIA, Energy Information Administration; UNSD, United Nations Statistics Division; FFDAS, Fossil Fuel Data Assimilation System; EDGAR, Emissions Database for Global Atmospheric Research; CDIAC, Carbon Dioxide Information Analysis Center; UNFCCC, United Nations Framework Convention on Climate Change; CAIT, Climate Analysis Indicators Tool; CEDS, Community Emissions Data System; GCP, Global Carbon Project; PRIMAP-hist, Potsdam Real-time Integrated Model for probabilistic Assessment of emissions Paths; CM, Carbon Monitor; GridFED, Gridded Fossil Emissions Dataset; GRACED, Global Gridded Daily CO_2 Emissions Dataset; ODIAC, Open-source Data Inventory for Anthropogenic CO_2. SOURCE: Modified from Andrew, 2020.

Administration, International Energy Agency, United Nations Statistics Division, and BP) in turn derive, in whole or part, energy consumption statistics from national sources in the form of questionnaires or surveys (Andres et al., 2012; Andrew, 2020). On the one hand, these efforts provide global data in a consistent manner; on the other hand, many of the energy data lack transparency due to confidentiality or intellectual property restrictions. This is a critical element in understanding the content of global/national inventories and in considering what independent information might be considered to evaluate and validate outcomes from the many global/national GHG inventories that use these source data. Primary inventories have been compiled globally that may combine two or more of the energy datasets to generate estimates of GHG emissions. There are a number of "mixed secondary/tertiary" inventories that use a mixture of primary energy/emissions data but also draw critical information from primary inventories. Finally, tertiary GHG inventories are primarily derived from primary inventories and typically provide additional space/time resolution (mostly through downscaling) or specific data additions.

It is important to note that the datasets listed in Figure 2-4 and Table 2-1 do not represent an exhaustive accounting of all global energy-related inventories. The goal of Figure 2-4 and Table 2-1 is to trace the relationship between national energy datasets and global inventories that utilize energy (and in some cases GHG emissions) information from those sources. While Figure 2-4 was modified from Andrew (2020), it does not adopt the same categories. This figure does not capture the details of individual inventories (e.g., timeliness of energy data updates), and the spatial resolution of each inventory is noted in Table 2-1.

For non-CO_2 emissions, in addition to the sources listed in Table 2-1, the U.S. Environmental Protection Agency (EPA) produces national inventories using IPCC default methodologies, international statistics for activity data, and IPCC Tier 1 default emission factors (EPA, 2019). Methane (CH_4) and nitrous oxide (N_2O) specific inventories have also been developed at the global and national scale using activity-based approaches for projections of mitigation potential (e.g., Harmsen et al., 2019; Höglund-Isaksson et al., 2020). The Food and Agriculture Organization of the United Nations is a key source of activity data for land use and agricultural emissions.

Because CO_2 emissions from the land use, land-use change, and forestry (LULUCF) sector use singular data and methods, they are often not incorporated into most of the global/national GHG inventories except in a few instances (e.g., Global Carbon Project, an annual summary report with their own gridded inventory; UNFCCC). However, several inventories specific to LULUCF sector emissions have been created (Table 2-1). Key distinctions among these global/national GHG inventories are whether the dataset includes carbonate sources (e.g., cement calcining), non-CO_2 GHGs, net anthropogenic biosphere CO_2, or industrial process/fugitive emissions (Table 2-1). Furthermore, some inventories separate agriculture and LULUCF while others combine them into a single sector (agriculture, forestry, and other land use [AFOLU]), making comparability challenging. There are also distinctions in how the information output is categorized and reported (e.g., whether the output is at subnational or subannual scales, the resolution, latency, amount of sector detail). Finally, there are distinctions in methodologies associated with emission factors and/or modeling differences (e.g., for net anthropogenic biosphere CO_2).

The energy-related global GHG inventories listed in Table 2-1 all reflect a Scope 1 or territorial accounting perspective. However, global GHG inventories that reflect a Scope 3 or consumption perspective have been developed, though this has been primarily at country resolution (e.g., Davis and Caldeira, 2010; Hertwich and Peters, 2009; Kanemoto et al., 2016; Moran et al., 2018). Consumption-based GHG emissions are included in the inventory development of the Global Carbon Project (Friedlingstein et al., 2022).

TABLE 2-1 National or Global Greenhouse Gas Inventories and Key Characteristics

Inventory Name	Gases	Geographic coverage	Resolution (space, time)	Time period (latency)	Sectors (yes/no)	Cement (yes/no)	Fuel type (yes/no)	Reference	Webpage
PRIMARY ENERGY DATASETS									
International Energy Agency (IEA)	Energy, CO_2, CH_4	190 countries	Country, annual	1971–2020 1960–2020 (OECD)	Yes	No	Yes	IEA, 2021a,b	https://www.iea.org/data-and-statistics/data-products
Energy Information Administration (EIA)	Energy, CO_2	230 countries	Country, annual	1980–2019 (country) 1949–2018 (global)	No	No	Yes	EIA, 2021	https://www.eia.gov/environment/emissions/state/
BP	Energy, CO_2	80 individual countries, 12 regional aggregates	Country/region, annual	1965–2021	No	Yes	Yes (energy), No (emissions)	BP, 2022	https://www.bp.com/en/global/corporate/energy-economics/statistical-review-of-world-energy.html
UN Statistics (UNSD)	Energy, CO_2	230 countries/territories	Country, annual	1990–2020 (publicly available)	No	No	Yes	Pachauri et al., 2014	https://unstats.un.org/unsd/energystats/data/ open data: http://data.un.org/Explorer.aspx
PRIMARY INVENTORIES									
United Nations Framework Convention on Climate Change (UNFCCC)	CO_2, CH_4, N_2O, CO, NMVOC, SO_2, HFCs, PFCs, SF_6, NF_3, CO_2-LULUCF	All Parties	Country, Annex I: annual, non-Annex I: biennial	1990–2020 (T-2) (developed) or 1990–various (T-4) (developing)	Yes	Yes	Yes	UNFCCC, 2021a	https://unfccc.int/process-and-meetings/transparency-and-reporting/greenhouse-gas-data/ghg-data-unfccc/ghg-data-from-unfccc
Emissions Database for Global Atmospheric Research (EDGAR)	CO_2, CH_4, N_2O, F-gases (HFC, PFCs, SF_6, NF_3), Air pollutants (SO_2, NO_X, NMVOC, CO, NH_3, $PM_{2.5}$, PM_{10}, BC, OC)	228 countries, global	0.1°×0.1°, monthly	1970–2018 (NA)	Yes	Yes	No	Crippa et al., 2021	https://edgar.jrc.ec.europa.eu/report_2021

Name	Gases	Coverage	Resolution	Time period				Reference	URL
Carbon Dioxide Information Analysis Center (CDIAC)	CO_2	259 countries	Country, annual	1751–2017	No	Yes	Yes	Gilfillan and Marland, 2021	https://cdiac.ess-dive.lbl.gov/
Gridded Global Model of City Footprints (GGM-CF)	CO_2 (consumption-based footprint)	13,000 cities	Urban area	2013	No	Yes	No	Moran et al., 2018	https://www.citycarbonfootprints.info/
MIXED SECONDARY/TERTIARY INVENTORIES									
Global Carbon Project (GCP)	CO_2, CO_2-LULUCF (also consumption-based CO_2)	259 countries ($FFCO_2$ only)	Country, annual	1750–2020 ($FFCO_2$) 1850–2020 (LULUCF)	Yes	Yes	Yes	Friedlingstein et al., 2022	https://www.globalcarbonproject.org/carbonbudget/
Fossil Fuel Data Assimilation System (FFDAS)	CO_2	137 countries, 3 regional aggregates	0.1°×0.1°, monthly	1997–2015	No	No	No	Asefi-Najafabady et al., 2014	https://ffdas.rc.nau.edu/Data.html
Community Emissions Data System (CEDS)	SO_2, NO_2, BC, OC, NH_3, NMVOC, CO, CO_2, CH_4, N_2O	221 countries	0.5°× 0.5°, monthly	1750–2019 1970–2019 (CH_4 and N_2O)	Yes	Yes	No (only global)	Hoesly and Smith, 2018; McDuffie et al., 2020; O'Rourke et al., 2021	http://www.globalchange.umd.edu/ceds/
Potsdam Real-time Integrated Model for probabilistic Assessment of emissions Paths (Primap-Hist)	CO_2, CH_4, N_2O, F-gases, HFCs, PFCs, SF_6, NF_3	All UNFCCC member states, most non-UNFCCC territories	Country, annual	1750–2019	Limited	Yes	No	Gütschow et al., 2019	https://www.pik-potsdam.de/paris-reality-check/primap-hist/
CAIT/Climate Watch	CO_2, CH_4, F-gases, N_2O	185 countries	Country, annual	1850–2019 (country) 1990–2019 (sectors)	Yes			Hennig et al., 2017	https://www.climatewatchdata.org/

continued

TABLE 2-1 Continued

Inventory Name	Gases	Geographic coverage	Resolution (space, time)	Time period (latency)	Sectors (yes/no)	Cement (yes/no)	Fuel type (yes/no)	Reference	Webpage
Open-source Data Inventory for Anthropogenic CO_2 (ODIAC)	CO_2	259 countries	1km×1km, monthly	2000–2019	Partial	Yes	Yes	Oda et al., 2018	https://db.cger.nies.go.jp/dataset/ODIAC/
GCP-GridFED	CO_2	259 countries	0.1°×0.1°, annual	1959–2018	Yes	Yes	Yes	Jones et al., 2021	https://mattwjones.co.uk/co2-emissions-gridded/
TERTIARY EMISSIONS									
Carbon Monitor	CO_2	6 countries, EU+UK, ROW	Country and region, 0.1°×0.1°, daily	2019–2022	Yes	Yes	No	Liu et al., 2020a	https://carbonmonitor.org/
Global Gridded Daily CO_2 Emissions Dataset (GRACED)	CO_2	6 countries, EU+UK, ROW	0.1°×0.1°, daily	2019–2020	Yes	Yes		Dou et al., 2022	
LULUCF/AFOLU									
FAOSTAT	CO_2, CH_4, N_2O, CO_2-LULUCF			1961–2019	Yes	NA	NA	Tubiello et al., 2013	https://www.fao.org/faostat/en/
BLUE	CO_2-LULUCF	Global	0.25°×0.25°, annual	1700–2019	No	NA	NA	Hansis et al., 2015; updated simulations described by Friedlingstein et al., 2020	
OSCAR	CO_2-LULUCF	10 regions	Regions, annual	1701–2019	No	NA	NA	Gasser et al., 2017; 2020; Friedlingstein et al., 2020	https://www.icos-cp.eu/science-and-impact/global-carbon-budget/2020

| H&N | CO_2-LULUCF | 187 countries | Country, annual | 1850–2019 | No | NA | NA | Friedlingstein et al., 2020; Houghton and Nassikas, 2017 |

NOTES: The efforts presented here attempt to capture the most established, commonly used greenhouse gas (GHG) inventories published in the peer-reviewed literature. It is not exhaustive of all global GHG inventories, particularly those under development, and contains a mixture of gridded and non-gridded inventories.
CO_2, carbon dioxide; CH_4, methane; N_2O, nitrous oxide; CO, carbon monoxide; NMVOC, non-methane volatile organic compound; SO_2, sulfur dioxide; F-gases, fluorinated gases; HFCs, hydrofluorocarbons; PFCs, perfluorochemicals; SF_6, sulfur hexafluoride; NF_3, nitrogen trifluoride; BC, black carbon; OC, organic carbon; NO_2, nitrogen dioxide; NO_x, nitrogen oxides; PM, particulate matter; NH_3, ammonia; $FFCO_2$, fossil fuel CO_2; CO_2-LULUCF, CO_2 from land use, land-use change, and forestry; LULUCF, land use, land-use change, and forestry; AFOLU, agriculture, forestry, and other land use; T, current year; EU, European Union; UK, United Kingdom; ROW, rest of world; OECD, Organisation for Economic Co-operation and Development.
SOURCE: Modified from Andrew, 2020 and Minx et al., 2021.

Regional/national GHG Inventories

Here we consider a separate scale of GHG inventories that span regional (multistate/province) to subnational (~provincial/state scale—larger than the local/urban scale) to national domains. They are relevant at the global scale as they provide some GHG inventory examples with features and innovations that are worth considering when evaluating the strengths and weaknesses of the global/national GHG inventories. We provide examples of these efforts in Box 2-2 and below to illuminate the approaches, innovations, and opportunities they offer.

China has quantified emissions in multiple ways, with subnational granularity, unique datasets, and detailed sector composition (e.g., Guan et al., 2012; Liu et al., 2015; Shan et al., 2020) (see also Case Study 4.2). The Multiresolution Emission Inventory for China—High Resolution (MEIC-HR) emissions data product merges county/provincial resolution nonpoint sources with a detailed database of factory and power plant emissions sources; point sources are quantified from facility-scale fuel consumption statistics (Zheng et al., 2021). A comprehensive comparison of different activity-based inventories has been performed by Han et al. (2020).

Notable in the United States and the United Kingdom are efforts to quantify consumption-based emissions at subnational scales (Jones and Kammen, 2015; Minx et al., 2013). Such efforts rely on economic consumption statistics and downscaling through divisions of economic activity into commodity categories. There is also a wider array of efforts aimed at individual process or individual GHGs at the regional/national scale. For example, Maasakkers et al. (2016) produced a methane emissions inventory for the United States based on the combination of numerous datasets. Similarly, over a dozen activity-based studies were conducted in the United States across the natural gas supply chain employing ground-based methods such as the use of tracer-flux, direct measurements such as high-flow sampling, calibrated bags, and mobile measurement techniques (Allen et al., 2013; Brantley et al., 2014; Lamb et al., 2015; Mitchell et al., 2015; Subramanian et al., 2015; von Fischer et al., 2017). The EPA also produces global non-CO_2 emissions inventories at the national scale in approximately annual reports (e.g., EPA, 2019, 2022).

BOX 2-2
The Vulcan Project

The Vulcan Project in the United States (Gurney et al., 2009, 2020b) quantifies fossil fuel CO_2 at fine space/time resolution. The Vulcan approach works from fuel-consuming assets (e.g., power plant, road segment, county buildings) and quantifies fossil fuel CO_2 emissions at the point of combustion. The resolution of the data product is point/line/polygon but it is also placed onto regular gridded (currently 1 km^2) formats for use in atmospheric modeling studies (Basu et al., 2020). The output is produced hourly and spans a multiyear timeframe. The underlying resolution in space/time and the functional attributes (e.g., building type, engine size) are key considerations in GHG inventory development as the mitigation decisions increasingly focus on sector- and subsector-specific mitigation policies.

Local/urban GHG Inventories

Considering the local/urban scale in an evaluation framework for GHG inventories is relevant because efforts at these scales demonstrate valuable attributes—including granularity and completeness—that make them more usable and relevant for decision makers. Approaches used at the local/urban scale could be scaled to larger jurisdictions and thereby enrich the suite of techniques at the global/national scale.

> **BOX 2-3**
> **Activity-based Approaches at City Scales: Global Covenant of Mayors**
>
> Approximately 70 percent of global greenhouse gases (GHGs) are emitted by cities. While responsible for the consumption of over two-thirds of the world's energy, cities are particularly vulnerable to the impacts of climate change (Reckien et al., 2018). As initiatives by cities to quantify and monitor GHGs for mitigation purposes have advanced, their climate targets have occasionally surpassed country-scale ambitions (Bertoldi et al., 2018).
>
> Regions have also identified knowledge gaps between various city/urban-scale inventories: differences in accounting methodologies, reference scenarios, and an absence of a harmonized datasets (Kona et al., 2018; Pachauri et al., 2014). This finding spurred the Global Covenant of Mayors' (GCoM) effort to create a harmonized, complete, and verified dataset of GHG inventories for 6,200 cities in European and Southern Mediterranean countries. Previous iterations of such data have proven to be extremely valuable for community engagement in climate action through policy and decision making (Pablo-Romero et al., 2018; Palermo et al., 2020). The improved accessibility and completeness of the dataset are expected to give rise to more studies and usage by scientific and nonscientific communities.
>
> Limitations and uncertainties in this effort primarily come from each city's capacity to gather and report data under the harmonized framework of the GCoM. There is limited information on the methods used by cities in determining their emissions, and uncertainties arise when cities use national/international-scale emission factors (Kona et al., 2021). Additionally, it is worth noting that validation is done through a feedback-rechecking system between experts and inventory developers. Finally, the GCoM EU cities are not required to report complete inventories. Instead, select emission sources that cities deem most relevant for management are reported, hindering comparability and other issues.

Local/urban GHG inventories are produced for a large portion of cities around the globe. For example, 9,120 cities representing over 770 million people (10.5% of the global population) have committed to the Global Covenant of Mayors (GCoM) to promote and support action to combat climate change (GCoM, 2018) (Box 2-3). Over 90 large cities, as part of the C40 network, have similarly committed to mitigation actions with demonstrable progress. However, fewer than 10 percent of cities participating in the GCoM-EU Secretariat have reported sufficient inventories for progress tracking (Hsu et al., 2020).

As with the GHG inventories at larger scales, the development of local/urban activity-based GHG inventories is done by local/urban practitioners (e.g., city sustainability staff, NGOs, consultants) and the research community with overlap between the two groups. Many of the practitioner-based local/urban activity-based GHG inventories are compiled using tools and protocols developed by the NGO community (Fong et al., 2014; WRI/WBCSD, 2004), though they are often based on IPCC methodologies developed for the national UNFCCC reporting process. Attempts at harmonizing approaches and methods are often done in conjunction with networks of cities sharing tools and protocols (e.g., International Council for Local Environmental Initiatives, GCoM, and C40). Because of the complexity and resource demands necessitated by GHG inventory development, the quality associated with the desired attributes of a GHG inventory (e.g., transparency, completeness) are often lacking or vary widely (Box 2-3).

Facility Scale (Corporate Inventories and Accounting)

Due to voluntary or mandatory requirements, companies produce GHG inventories to assess the efficacy of corporate emissions mitigation strategies. Corporate GHG inventories often also support national or regional climate strategies (EPA, 2009; Klaaßen and Stoll, 2021; Marlowe and Clarke, 2022; OECD, 2015; WRI, 2015). As mentioned in Chapter 1, the Task Force for Climate-related Financial Disclosures is leading to new requirements for companies, especially in the financial services sector, to develop climate disclosure reporting and plans. At this point, over 40 countries have some amount of corporate regulated reporting requirements, including 15 of the G20 countries (OECD, 2015; WRI, 2015). The reporting of GHG emissions is considered a key corporate environmental, social, and governance (ESG) disclosure metric by a variety of stakeholders, including policy makers, shareholders, customers, employees, and the public. Such disclosures are made via stand-alone corporate responsibility reports, and in some cases in financial reports or other regulatory mechanisms.

The International Organization for Standardization (ISO) 14064-1 (ISO, 2018) provides an internationally recognized framework for quantification and reporting of GHG emissions inventories. The GHG Protocol Corporate Accounting and Reporting Standard has been widely used (Klaaßen and Stoll, 2021; SEC, 2022) for about two decades. Corporate GHG inventories estimate the GHG emissions associated with a company's operations, financial control, or equity-share of particular markets. Most regulatory reporting requirements focus on Scope 1 or direct emissions, but a few require Scope 2 reporting, and some encourage Scope 3 reporting as there is increasing interest in complete supply-chain emission impacts (OECD, 2015; WRI/WBCSD, 2011). The corporate GHG emissions accounting follows the activity-based approach described above (Equation 1). Besides the GHG Protocol Corporate Standard, there are industry-specific supplemental voluntary guidelines. Examples include the GHG Protocol Agricultural Guidance (Box 2-4) and the American Petroleum Institute Compendium for the oil and gas industry. Mandatory reporting regulations also may specify certain accounting methods and scope and reporting thresholds.

Over the past decade, the use of life-cycle analysis (LCA)—a cradle-to-grave or cradle-to-cradle analysis technique to assess environmental impacts associated with all the stages of a product's life—to estimate environmental impacts, including GHG emissions, of a product (e.g., fuel or technology) has grown. Unlike conventional inventories, LCA evaluates the impact of a product or a technology over its life cycle, normalizing the impacts of each stage of the supply chain to a defined functional unit (e.g., 1 megawatt-hour of electricity or 1 ton of natural gas).

LCA can be applied to domestic regulatory assessments. For example, LCA may be used under EU Taxonomy to screen natural gas-powered generation (European Commission, 2022) or to track

---

BOX 2-4
The Greenhouse Gas Protocol Agricultural Guidance

In 2014, the first global guidance to measure greenhouse gas (GHG) emissions from the agriculture sector was released. This guidance was motivated by agriculture's significant contribution to global emissions (approximately 17%) and lack of emissions reporting. World Resource Institute's GHG Protocol Agriculture Guidance (https://ghgprotocol.org/agriculture-guidance) is intended to promote emissions reporting of agricultural companies.

The agricultural guidance summarizes and customizes the Corporate Standard to the agricultural sector and provides additional recommendations specific to the sector. A key characteristic of agricultural emissions is that they typically only generate a subset of GHGs compared to other corporate sources. The guidance also provides recommendations on what emissions and removals should (Scope 3 sources, CO_2 fluxes from carbon stocks in soils, removals by woody vegetation) and should not (CO_2 removals by herbaceous vegetation) be reported.

and reduce the carbon intensity of transportation fuels in California and Oregon.[3] This approach places the environmental impacts across the supply chain relative to each other, better supporting corporate and public policy decision making by avoiding over-counting emissions as in Scope 3 methods when supply chains are not vertically integrated.

Atmospheric-Based Approaches

Atmospheric-based approaches (often referred to as "top-down" approaches) use atmospheric measurements and an understanding of atmospheric transport and chemical processes (in some cases) to infer information on emissions and removals (fluxes to and from the atmosphere). Atmospheric measurements can be either mixing ratio observations or flux observations, such as from an eddy flux measurement. The key distinction with atmospheric-based approaches is that atmospheric dynamics must be incorporated to estimate emissions or uptake for a specific space and time. In this section, we provide an overview of the range of analysis approaches used to transform atmospheric measurements to estimates of emissions or removals (e.g., emission ratios, source types, flux estimates, source locations) and the surface-, aircraft-, and space-based measurement techniques that are foundational to these approaches.

An important caveat to understand about these approaches is that many GHGs are comprised of both anthropogenic and natural sources. For gases like CO_2 and methane, atmospheric concentrations account for both human and natural fluxes, including emissions and removals from plants, soils, and the ocean for CO_2, and wetlands and seeps in the case of methane. As discussed briefly in later sections, this means that additional or more complex analysis is required to gain insights into the anthropogenic component when inferring fluxes from atmospheric data due to the large role of natural fluxes. Note that many fluorinated gases have no significant natural sources; as a result, atmospheric measurements can largely be directly interpreted as an anthropogenic release for those gases.

Mass Balance Approach

The mass balance approach is used to quantify GHG emissions at urban to regional scales. The approach has been used extensively to quantify point sources—for example, oil and gas facilities, power plants, and landfills (Cambaliza et al., 2017; Conley et al., 2017)—but has seen considerable application estimating city-scale GHG emissions. The approach involves high-precision measurements of GHG mixing ratios in the boundary layer (from the near surface to the top of the boundary layer) downwind of the urban environment and calculation of the net incremental outflow of GHGs from the city, region, or point source (e.g., Andrews, 2022; Heimburger et al., 2017), using appropriate measurements of wind speed and direction. The net outflow of GHGs is obtained by subtraction of the background inflow GHG concentrations, which is mostly determined through measurements of GHG concentrations at the extended edges of downwind (perpendicular to the mean wind direction) transects that do not receive outflow from the city. This is typically the most challenging aspect of the mass balance experiments, as the background can be quite uncertain compared to the incremental concentration. The incremental concentration at each point downwind (typically in an imaginary plane that extends from the surface through the top of the boundary layer, oriented perpendicular to the mean flow that captures the full horizontal and vertical extent of the plume) is multiplied by the perpendicular component of wind speed relative to the plume to calculate the flux through the perpendicular imaginary plane. The total emission rate—obtained by

---

[3] See California's Low Carbon Fuel Standard, https://ww2.arb.ca.gov/our-work/programs/low-carbon-fuel-standard.

integrating over all net flux points covering the entire plume—is then attributed to the integrated upwind surface emission sources. One notable limitation of the mass balance approach is that it usually provides only a snapshot and requires an extrapolation from the day of measurement to longer time scales. In addition, most mass balance approaches have equipment-level attributional limits since the flux rates are generated at a facility-level or higher spatial scale (Schwietzke et al., 2017). Mass balance approaches have also been applied to point sources measured by satellite data.

The calculated total region emission rate can be compared to an inventory. Such approaches have been used as part of the Indianapolis Flux Experiment project to assess CO_2 and CH_4 emissions from Indianapolis (see Case Study 4.3) (Cambaliza et al., 2014; Heimburger et al., 2017), and can be used to assess independent inventories or to calculate GHG emissions from cities, regions, or point sources that do not have emission inventories. The same sampling plans (e.g., aircraft flight plans) have also been used to enable inverse modeling calculations of emission rates (Lopez-Coto et al., 2020; Pitt et al., 2022). As with other atmospheric-based approaches, there may be a mixture of anthropogenic and natural sources within the region of interest (e.g., CO_2 fluxes from vegetation and soils within cities), which must be accounted for by using tracers or other methods to isolate anthropogenic emissions.

Isotopic Analysis and Multigas Approaches

Isotopic analysis and multigas approaches can be useful ways to identify sources of GHG emissions and, in some cases, estimate emissions. Isotopic analysis utilizes radiocarbon (^{14}C) measurements of atmospheric CO_2 to distinguish fossil fuel emissions (which do not contain radiocarbon atoms) from modern CO_2 fluxes (i.e., from wood burning). Essentially, isotopic analysis methods can be used to quantify the impact of the increasing burden of anthropogenic emissions on atmospheric CO_2 since the pre-industrial era (Basu et al., 2016, 2020). Similar approaches for methane measurements are currently being developed (Zazzeri et al., 2021).

Tracer-to-tracer (or multigas) approaches emerged because many GHG observing programs combine in situ CO_2 measurements with co-located emission tracers (e.g., non-methane volatile organic compounds, carbon monoxide [CO], nitrogen oxides, etc.). Analysis of correlations between CO_2 and the emission tracers can be used to attribute emission sources. For example, ethane and butane are associated with natural gas heating, nitrogen oxides with transportation and traffic, and sulfur dioxide with coal and diesel combustion. Multigas methods have been used to highlight inaccuracies in CO_2 emissions, tracer emission ratios in inventories, and have been used to estimate fugitive methane emissions from urban centers (Ammoura et al., 2016; Plant et al., 2019). One challenge of these approaches is that they sometimes require knowledge of atmospheric chemistry that is typically simplified to the detriment of the analysis.

Tracer-to-tracer observations have also been used to estimate emissions. By assuming that the emissions of two gases are reasonably co-located and emissions of one tracer gas are well known from activity-based approaches, the emissions of that gas are scaled with observed tracer ratios to estimate emissions of the gas of interest. For example, measurements of hydrofluorocarbons (HFCs) and methane to CO concentrations have been applied to estimate HFC and methane emissions by scaling CO emissions (Greally et al., 2007; Hsu et al., 2010). Emissions estimated using this technique can include biases from incorrect assumptions, however. More details on isotopic analysis and multigas approaches can be found in Appendix B.

Inverse Approach

Inverse modeling in the context of GHG emissions estimation refers to an approach that starts with observations of atmospheric mixing ratios and infers GHG emissions by accounting for the atmospheric mixing and transport (see Figure 2-5). There are many variations of the general inverse approach that consider additional information constraints, the optimization procedure, and different conceptions of atmospheric transport modeling. Activity-based inventories, described above, are often used as inputs to inversion analyses.

The general inverse approach attempts to find the set of scaling values that transform surface fluxes (potentially including anthropogenic and natural emissions and removals) in the geography of interest, such that after transport in an atmospheric model, there is an optimal (minimum distance) match to observed atmospheric mixing ratios. Because atmospheric observations are an insufficient constraint except in extreme cases, a prior estimate of emissions and removals (e.g., from an activity-based approach) and their associated uncertainties is required. This type of inverse modeling is Bayesian in that it combines observations with prior knowledge to produce an estimate of emissions (Enting, 2001; Tarantola, 1987). This method is the most common type of inverse modeling at global or regional scales.

In a typical application, an atmospheric transport model is used to simulate the transport of GHGs given a certain spatial distribution of fluxes. These fluxes are termed the prior estimate and are constructed as described in the activity-based section above. Then, comparing simulated concentrations with those from ground-based, aircraft, or satellite observations, the fluxes in individual grid cells or larger regions are scaled to improve the fit to the observations. The resulting estimate of emissions is termed the posterior estimate. Uncertainties in the prior estimate and in the observations, as well as in the transport model, must be specified, and uncertainties in the posterior are also calculated. Specific techniques include the traditional synthesis inversion that is applied for a

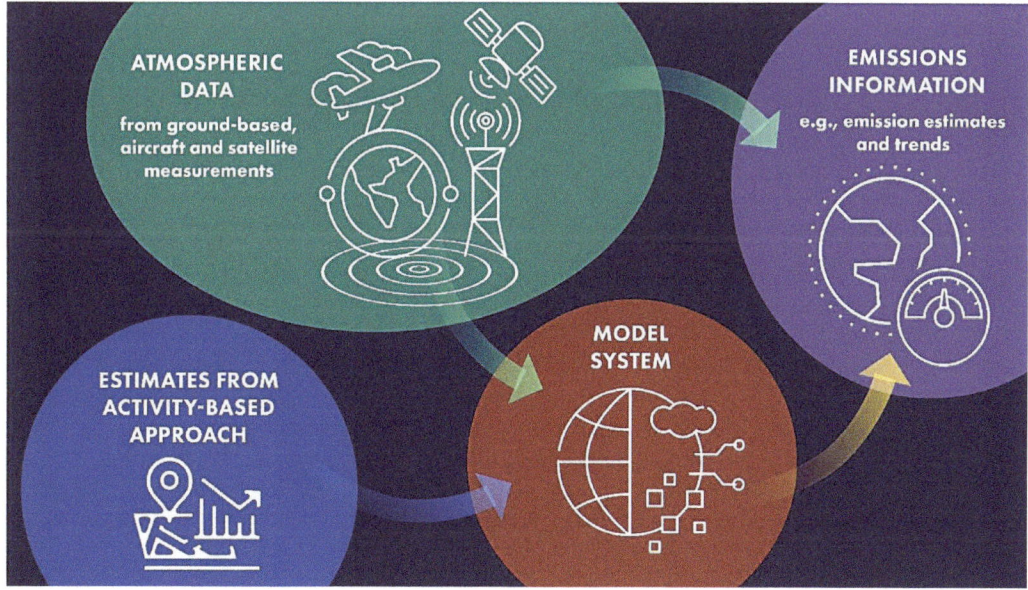

FIGURE 2-5 Atmospheric-based approaches use atmospheric measurements (green) together with information about atmospheric transport and chemical processes from observations and/or from a model system (red) and/or other prior information from an activity-based inventory (blue) to infer greenhouse gas emissions (purple).

limited number of regions, as well as adjoint, Kalman filter, 4-DVar (four-dimensional variational data assimilation), and hierarchical estimation methods that can be applied for higher spatial and temporal resolution in posterior estimates using large datasets.

Bayesian inverse modeling is quantitative but it requires decisions to be made by the user about the setup of the calculation, and the results can be influenced by several factors, including errors in the atmospheric transport model and the spatial and temporal resolution of the prior estimate of emissions (Manning, 2011). To address some of these factors, often a number of sensitivity tests are run (e.g., with different transport models or prior estimates) (Bergamaschi et al., 2018) to identify robust features in posterior emissions estimates. A common challenge is specifying the prior uncertainty and particularly the prior uncertainty correlations between different regions or grid cells, which are sometimes ignored or assumed to have a spatial correlation that exponentially decays with distance (Rödenbeck et al., 2003).

Another challenge with Bayesian inverse modeling is a lack of constraints on source attribution. Often the method is employed assuming the spatial distribution of the emissions from various sources is correct in the prior estimate so that biases in the prior model are carried through to the posterior estimate. Adding other tracers to the inversion or using them to isolate a particular sector before inversion, for example by using radiocarbon measurements to isolate fossil fuel CO_2 (Basu et al., 2020; Graven et al., 2018), is one way to apply additional constraints for source attribution. For some regions, even in some megacities, the magnitude of natural emissions and removals is large, making the quantification of solely anthropogenic emissions challenging (e.g., Miller et al., 2020). In conventional atmospheric CO_2 inversions (e.g., Crowell et al., 2019; Gurney et al., 2002; Peylin et al., 2013), fossil fuel emissions are typically assumed to be specified perfectly by prior inventories so that biospheric CO_2 fluxes (natural and anthropogenic) can be estimated. Recent comparisons of large-scale CO_2 inversion results to activity-based inventories for anthropogenic fluxes have focused on the AFOLU sector (e.g., Byrne et al., 2022; Deng et al., 2022).

Observations Utilized in Atmospheric-Based Approaches

Atmospheric observations from surface-, aircraft-, and satellite-based measurements have strengths and weaknesses (e.g., coverage, accuracy and precision, spatiotemporal resolution) for constraining GHG emissions via atmospheric-based approaches. In the section that follows, we briefly introduce the range of atmospheric observations utilized in the approaches described above including GHGs, co-emitted gases, and isotopes. Detailed information about these measurements can be found in Appendix B.

Surface-based observations. In situ continuous and flask measurements of atmospheric GHGs have been used to improve understanding of the variability and trends of emissions and carbon sinks at different scales (see Box 2-5). The global surface network monitors GHGs in remote locations and the measurements have been used, for example, to quantify the increase of atmospheric GHG concentrations over the past decades—key information for building climate projection scenarios (e.g., Meinshausen et al., 2017)—and to constrain global inversion models (e.g., Bousquet et al., 2000; Denning et al., 1995; Fernández-Martínez et al., 2019; Rayner et al., 2008) and assimilation systems such as CarbonTracker,[4] the Copernicus Atmosphere Monitoring Service (e.g., Pinty et al., 2019), and the National Aeronautics and Space Administration's (NASA's) Goddard Earth Observing System (GEOS) (Weir et al., 2021).

Continental networks monitor atmospheric GHGs in the lower atmosphere and these GHG measurements have been used to assess trends and variability of atmospheric GHG concentrations, sources, and sinks at regional to continental scales (see Appendix B). Continental GHG networks have been mostly developed in Europe and the United States, and some parts of the world are still

---

[4] https://gml.noaa.gov/ccgg/carbontracker

> **BOX 2-5**
> **Atmospheric Measurement Requirements**
>
> The atmospheric greenhouse gas (GHG) concentration growth rate and trends can be inferred from atmospheric measurements, provided they are accurate and precise. Scientific calibration of data is a critical piece of GHG monitoring for data that are used for quantitative analysis. Calibration of data is crucial for intercomparison and synthesis of data from different instruments and sources. Continuous and reliable measurements are ensured by the Global Atmosphere Watch (GAW) program of the World Meteorological Organization (WMO). The data are available from the World Data Centre for Greenhouse Gases (WDCGG) at the Japan Meteorological Agency.[a] WMO/GAW delivers recommendations for the accuracy and precision needed for different GHG measurements, and an international calibration scale for the gas standards deployed at the observation sites. Detailed recommendations on GHGs, isotopes, co-emitted species, and ancillary measurements can be found in WMO/IAEA (2013).
>
> Influences on GHG concentrations from regional, urban, or point-source emissions or sinks can be measured by in situ, flask, or remote sensing measurements and used in the different atmospheric approaches presented in this section. Assessment of the effect of emissions or sinks requires monitoring "background" GHG concentrations (i.e., local concentrations not affected by emissions). Background concentrations can be measured through mobile platforms or fixed observation sites located upwind of the plume. Then, at least one additional site is needed to collect measurements downwind of emissions. Both datasets must be related to the same calibration scale to be comparable. At GHG observation sites, WMO/GAW recommends also monitoring ancillary measurements such as meteorological parameters and the height of the boundary layer—the first layer of the atmosphere into which GHGs are emitted, mixed, and transported.
>
> Atmospheric-based approaches often rely on additional measurements of species that are co-emitted with GHGs (see Appendix B). Some co-emitted species are monitored continuously by air quality agencies or academic research groups, especially in urban and industrial areas. There are large potential co-benefits for both the GHG and air quality communities to work more closely as some GHGs and air pollutants that impact both climate and air quality have common sources.
>
> ---
> [a] https://gaw.kishou.go.jp/

underequipped, especially in the Southern Hemisphere (e.g., Africa, South America). The European ICOS-Atmosphere (Integrated Carbon Observing System[5]) datasets feed the European Copernicus GHG assimilation system[6] and the U.S. datasets are used as primary inputs for National Oceanic and Atmospheric Administration's (NOAA's) CarbonTracker CO_2 and methane data assimilation systems.

Urban GHG networks have been used to better quantify emissions and understand uncertainty and GHG concentration variability at urban scales (e.g., Xueref-Remy et al., 2018). Sensor networks have been used to quantify emissions from major urban sources (i.e., building, transportation, and industrial sectors) and more broadly to characterize GHG emission changes and compare process models to observations (e.g., Fitzmaurice et al., 2022; Kim et al., 2022a; Shusterman et al., 2016; Turner et al., 2020). Additional measurements of emission tracers at these scales can be tools to discriminate emission sources (e.g., Ammoura et al., 2016; Graven et al., 2018). Emerging low-cost sensors have been tested in recent years in some urban networks, for example the Berkeley Environmental Air-quality & CO_2 Network (BEACO2N) in the San Francisco Bay Area (Kim et al., 2022b; Turner et al., 2020) and the Carbosense network in Switzerland (Müller et al.,

---
[5] https://www.icos-cp.eu/
[6] https://atmosphere.copernicus.eu/ghg-services

2020). Such sensors were shown to be able to detect large urban plumes but need to be calibrated carefully and often.

Eddy covariance (EC) measures GHG fluxes (mostly CO_2 and methane) from high-frequency GHG concentrations and wind measurements. EC measurements have provided estimates of GHG emissions at city centers to independently assess emission inventories. For example, in London, Helfter et al. (2016) found a good agreement between the local inventory and EC emission estimates for CO_2 and CO, but a twofold higher emission of methane by the EC approach than in the inventory.

Ground-based remote sensing measurements have been used to characterize atmospheric GHGs at the facility scale, and to quantify GHG emissions at the urban scale (e.g., Wunch et al., 2009), and they are critical to the validation of satellite measurements. Several continuous emissions monitoring technologies have now been deployed at oil and gas sites that allow detection of methane emissions, and are in early stages of quantification (Alden et al., 2019; Chen et al., 2022a; Mingle, 2019). Ground-based remote sensing technologies may play a larger role in the coming years (see Appendix B). For example, the St. Petersburg experiment demonstrated the utility of these measurements to constrain urban CO_2 emissions (Ionov et al., 2021).

Aircraft-based observations. Aircraft measurements have been used for several decades to address the vertical variability of atmospheric GHGs (e.g., Machida et al., 2008; Xueref-Remy et al., 2011b), better understand GHG transport processes (e.g., Schuck et al., 2009), evaluate models, and validate satellite observations (e.g., Frankenberg et al., 2016). At the regional scale, airborne campaigns have provided critical data to constrain inverse modeling frameworks (e.g., Chen et al., 2019; Xueref-Remy et al., 2011a, 2011b). At the urban scale, airborne measurements have been used to quantify urban emissions through mass balance approaches (e.g., Cambaliza et al., 2015). Aircraft remote sensing measurements (not collecting samples into containers but measuring reflected sunlight or some other light signal) have been used widely for CO_2 and methane measurements (see Case Study 4.4). These techniques are typically used at a facility or local scale and may be done using a mass balance approach or plume seeking approach (NASEM, 2018).

Space-based observations. Satellite instruments GOSAT (Greenhouse gases Observing SATellite), GOSAT-2, OCO-2 (Orbiting Carbon Observatory), OCO-3, and TanSat (i.e., CarbonSat) have collected global measurements of total column atmospheric CO_2 from space. GOSAT and GOSAT-2 also measure column methane while TROPOMI (TROPOspheric Monitoring Instrument) measures methane but not CO_2. Satellite data have been used to quantify emissions from point sources, observe enhancements from cities, and estimate regional fluxes (CEOS, 2018; UNFCCC, 2022b). Satellite observations of tracers such as CO or nitrous oxides (NO_2) have also been used to investigate GHG emissions alone or in conjunction with satellite CO_2 and methane measurements.

Signal-to-noise ratios for total column CO_2 and methane can be low, making detection of emissions challenging. CO_2 is very well mixed, with global gradients and regional enhancements of a few parts per million (ppm) relative to a background of just over 400 ppm. To enhance the ability of satellite observations to capture anthropogenic CO_2, NO_2 measurements have been used (e.g., Hakkarainen et al., 2021; Reuter et al., 2019). Next generation satellites including GOSAT-GW (Greenhouse gases and Water cycle) and CO2M (Copernicus CO_2 Monitoring Mission) will collect NO_2 measurements in addition to CO_2 and methane. For methane, background concentrations are lower and discrete sources can create more easily detected enhancements; however, there is still a limit to the detection of concentration gradients caused by regional fluxes or point sources. Cloud cover, aerosols, and surface characteristics are important factors affecting the measurement coverage and precision possible with satellite measurements.

Satellite measurements have evolved from large to increasingly smaller footprints, improving the precision and spatiotemporal information available for analysis. For methane, there have been rapid advances in sensitivity and reduced footprints, with participation from governments, nonprofits, and the private sector. The planned next generation of CO_2 satellite measurements is projected to increase the coverage of measurements to nearly global each day. There are also new approaches that are designed to observe individual plumes, some with targeting, such as GHGSat, CarbonMapper (planned 2023 launch), and MethaneSAT with a wide-swath, small footprint (2 km^2) (planned launch 2022) (see Appendix B). Smaller footprints, wide-swath measurements, lower detection thresholds, and delivery of flux products along with concentration data may significantly increase the value of satellite measurements in the development of GHG emission inventories.

A large majority of satellite measurements originate from federal-level agencies, such as NASA in the United States, European Space Agency in Europe, and Japan Aerospace Exploration Agency and National Institute for Environmental Studies in Japan. These agencies historically have provided at least the higher-level science data products (retrieved atmospheric concentrations) freely and openly, and, in recent years, more data have been freely shared, including the fundamental measurements such as reflected light levels. Technology advances have resulted in rapid growth in the number of private/commercial satellites for GHG measurements (e.g., GHG-Sat, MethaneSAT, CarbonMapper). Data sharing policies vary; some datasets are not freely available, as income from data sales is part of the business plan for these private/commercial missions, which poses a risk to transparency (see Chapter 1).

Example Applications of Atmospheric-Based Approaches Across Scales

There are many example applications of atmospheric-based approaches, including the use of surface, aircraft, and satellite measurements, at all scales—global, national, regional, city, and facility. Most of the work that has been tied to emission inventories and policy-relevant scales has been at the national level and smaller. A few of these examples are highlighted here.

Using in situ measurements, the emissions of HFCs in the United Kingdom (U.K.) were estimated and compared to activity-based estimates (Manning et al., 2021). Data from six in situ measurement stations were used in the Inversion Technique for Emission Modelling (InTEM) (Arnold et al., 2018; Manning et al., 2011), and annual emissions estimates were derived for 2013 through 2020 for the 10 HFC gases that are reported to the UNFCCC. This work showed that U.K. HFC emissions had declined 35 percent in 2020 relative to the 2009–2012 period, and that the atmospheric observation-based estimates were on average 73 percent of the total HFC emissions from the U.K. GHG inventory. A similar atmospheric-based approach is used to estimate U.K. emissions of non-CO_2 GHGs on an annual basis, and these estimates are included in their national reporting to the UNFCCC; the United Kingdom is one of only a few countries to include atmospheric-based estimates in UNFCCC reporting.

Another national-scale example fundamentally driven by atmospheric data is the Basu et al. (2020) analysis of CO_2 emissions from fossil fuel combustion and cement production in the United States. This is the only atmospheric-based estimate of national fossil fuel CO_2 emissions available to date, even though fossil fuel CO_2 emissions are responsible for the majority of total GHG emissions. It relied on both CO_2 and the radiocarbon in CO_2 (^{14}C) to specifically evaluate the fossil fuel and cement emissions. They performed three inversions, using the variational framework of Basu et al. (2016) and three different gridded prior estimates. Their estimate of total U.S. national emissions of fossil fuel CO_2 is consistent with the national estimate made by the EPA, but it is significantly larger than some other activity-based estimates. The estimate they derived is also consistent with the U.S.-specific, high-resolution Vulcan 3.0 activity-based estimate. This work demonstrated the potential of a U.S.-based inversion system and the critical need for $\Delta^{14}C$ (or other tracer) measurements to isolate fossil fuel CO_2 emissions using atmospheric-based approaches.

In another example, Henne et al. (2016) used in situ measurements of methane in an inversion framework for Switzerland, finding good agreement with the total national emissions of the Swiss Greenhouse Gas Inventory, but some differences in the spatial distribution in their results relative to the national inventory. Switzerland is one of the countries other than the United Kingdom that has included atmospheric-based non-CO_2 GHG emissions estimates in UNFCCC reporting. Other efforts have used high-resolution inverse modeling to evaluate reported methane emissions for several countries (e.g., Petrescu et al., 2021; Wang et al., 2019; Worden et al., 2022) (see Case Study 4.5 about the VERIFY project).

Many studies have also applied these techniques at subnational scales. A large amount of research has been conducted on methane emissions in the U.S. state of California, including towers, aircraft, and satellites. Atmospheric inverse studies have indicated that methane emissions are higher than activity-based inventories developed by the state of California (Cui et al., 2017, 2019). Other activities evaluated the contribution of large point sources and various source types (e.g., Duren et al., 2019) such as agriculture (e.g., Amini et al., 2022; Heerah et al., 2021), gas distribution (e.g., He et al., 2019), and oil and gas extraction (e.g., Zhou et al., 2021) to California methane emissions and the differences between atmospheric- and activity-based estimates. Identifying point sources can be important as they may be easy targets for mitigation through emissions reductions, although approaches that focus on point sources may miss diffuse sources that are potentially equally or more important.

Point source emissions have also been estimated in other locations with aircraft and satellite data for methane (Cusworth et al., 2021a,b; Krautwurst et al., 2021; Varon et al., 2018) and CO_2 (Cusworth et al., 2021a,b; Nassar et al., 2017; Reuter et al., 2019). Atmospheric approaches may be particularly useful for investigating unmonitored sources and the identification of anomalous or upset conditions. While many of these datasets are publicly available, data and methods from private companies may not be published or made available for the research community to evaluate and reproduce.

Hybrid Approaches

Hybrid approaches derive GHG emissions information through the combination and more complete integration within and between the different approaches described thus far (Figure 2-6). Hybrid approaches are just starting to be developed and there are presently few examples, but many potential directions for further development. For example, a deeper integration of the atmospheric- and activity-based approaches, beyond what is practiced in the current suite of inverse studies, might functionally combine emission processes and atmospheric transport modeling such that atmospheric observations are directly adjusting emission factors or activity data. Another example is an activity-based approach with multiple overlapping core datasets further constrained by atmospheric-based mass balance estimates. Hybrid approaches may also take advantage of a range of new or nontraditional data sources combined with machine learning (ML) techniques that can efficiently analyze large datasets and derive relationships between activity, infrastructure, and GHG emissions in potentially new ways. The hybrid approach classification is less categorically distinct when compared to activity- or atmospheric-based approaches and represents a continuum of simple to more complex efforts aimed at greater integration and combination of the many measurement and modeling efforts to estimate GHG emissions and removals. This section provides an overview of existing and notional examples representing a trajectory toward more integration of existing approaches.

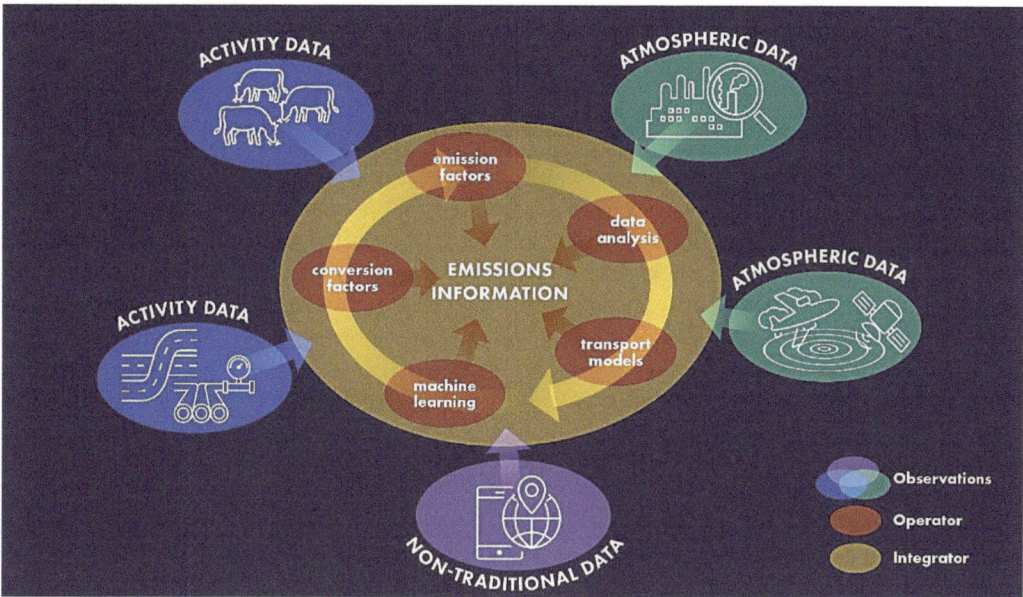

FIGURE 2-6 Hybrid approaches generate greenhouse gas emissions information through the combination and more complete integration of activity data (blue), non-traditional data (purple), and atmospheric data (green) that is modified by operator(s) (red) and integrated (gold).

Information Integration Via Data Assimilation

Data assimilation as practiced by numerical weather prediction is a useful model for a more integrated approach to quantifying GHG emissions. In contrast to atmospheric inversions already described, a data assimilation approach could expand the central model from an atmospheric transport model to a dynamical model that simulates GHG emissions and sinks and moves the fluxes through the atmosphere. In this way, the observational constraint moves beyond the prior flux and mixing ratio combination to a wider array of observed quantities, focusing on correcting and estimating the flux model parameter space instead of the fluxes (e.g., Kaminski et al., 2022; Rayner et al., 2005). Elements of activity data (e.g., continuous emission monitors, traffic counters, and gas meter data) in addition to atmospheric concentrations can be directly ingested into a core assimilation model offering degrees of constraint dictated by uncertainty. State estimation-based approaches (e.g., European Centre for Medium-Range Weather Forecasts) can quantify potential biases in activity-based estimates by comparing calculated and observed GHG concentrations, rather than emissions or fluxes (e.g., Weir et al., 2021). State estimation-based approaches are often based on an online meteorological or Earth system model, rather than an offline transport model, and can directly contribute to concentration monitoring as well as large-scale mitigation impact assessment in a more integrated manner (e.g., GHGs and air quality).

Other possibilities might integrate atmospheric-based measurements with other types of measurements not typically included in atmospheric inversions. For example, combining atmospheric measurements (e.g., aircrafts or drones) with novel continuous measurement technologies, especially for methane emissions detection, could help to establish the frequency and duration of emission events at a facility or larger basin level (e.g., Allen et al., 2022; Chen et al., 2022b; Wang et al., 2022b), and the impacts on regional emissions as a whole. Such approaches may hold potential for the development of empirical-based annualized inventory estimates, especially in the oil and gas sector.

Alternative Data Integration Through Machine Learning

Recent advances in ML—the application of computational algorithms usually applied to large-scale datasets to simulate human learning (Wang et al., 2009)—represent another potential hybrid GHG estimation approach to model complex, nonlinear, and nonparametric relationships between data to achieve potentially more complete and new GHG emissions information. ML algorithms and ML-powered models can take a series of data inputs to train a model to uncover statistical patterns, making predictions on new, "unseen" data (Huntingford et al., 2019; Milojevic-Dupont and Creutzig, 2021). These approaches are now being applied to massive datasets, incorporating multiple data types, and often integrated within emerging digital infrastructures (e.g., blockchains, distributed ledgers; see Box 2-6). While these approaches have the potential to address critical GHG emissions information gaps, including temporal and spatial resolution of existing data, they could undermine transparency needed for credible GHG accounting because ML algorithms are prone to be criticized as a "black box" due to their complexity and often difficulty in interpretation (Castelvecchi, 2016) (see Chapter 3).

To date, ML approaches have been applied in four broad categories related to Earth observation (EO): feature classification (e.g., land use or land cover and land cover change), anomaly, target and change detection, and regression-based methods (e.g., estimating a variable of interest such as GHG emissions from a set of underlying predictor variables) (Salcedo-Sanz et al., 2020). Since ML trains a computer to "learn" and identify relationships with data inputs, ML could improve the performance of traditional data fusion algorithms and approaches (Meng et al., 2020).

An example ML application to GHG emissions estimation is the use of ML-driven predictive models that take advantage of large EO datasets and high-resolution satellites that are capable of measuring land cover and atmospheric conditions. Given these two trends, there is active development of ML algorithms that attempt to estimate GHG emissions at multiple scales (i.e., national and subnational) given known emissions drivers—electricity generation, mobile transport, industry, deforestation, and land-use change (Akhshik et al., 2022; Hamrani et al., 2020; Han et al., 2021; Hsu et al., 2022). Two primary ML-based GHG estimation approaches exist: top-down methods that estimate emissions from socioeconomic and demographic drivers (from Hsu et al., 2022) and bottom-up approaches that characterize infrastructure or activity data rather than the emissions themselves. In the latter application, Climate TRACE has convened a consortium of more than 12 organizations developing ML-based approaches to estimate activity data (e.g., how frequently a power plant is in operation) and GHG emissions (Climate TRACE, n.d.) (see Case Study 4.6).

Numerous research studies are exploring ML-based approaches. Alova et al. (2021), for instance, developed an ML-based model to predict Africa's electricity generation mix in 2030 using asset-level data on current and planned power plants. Meanwhile, Han et al. (2021) utilized ML to estimate corporate GHG emissions data, incorporating more than 1,000 variables including industry sector classification data, revenue, corporate locations, and ESG data. Niu et al. (2020) applied neural network ML algorithms and random forests—an ML technique used to solve regression and classification problems—to determine 25 socioeconomic characteristics, such as gross domestic product growth rate, population, demographic structure, energy and industrial structure, energy intensity, and environmental policy stringency, and then used these variables to forecast scenarios of China's future emissions trajectory. While these recent studies provide some potential for ML approaches to address gaps and yield new insights and information for GHG emissions, to what degree they will improve upon existing activity- or atmospheric-based approaches is still uncertain.

> **BOX 2-6**
> **Distributed Ledger Technology for Data Storage**
>
> In addition to Earth observation data, digital ecosystems connecting "smart things" and scaled through internet infrastructure, referred to as the Internet of Things (IoT), promise to foster better data collection, integration, and analysis, allowing for improved services and diagnostics for environmental management (Chen et al., 2015; Tsai et al., 2014). Distributed ledger technologies that underpin blockchains have been discussed for their potential to provide decentralized, immutable records for climate data that can be integrated with smart contracts and oracles that can lower barriers to tracking and aggregating climate emissions data (Schletz et al., 2022). Blockchains also show promise in solving data privacy and trust issues associated with data sharing with zero-knowledge proofs, cryptographic tools that establish that sensitive data are available and verified but not visible to the public (Ben-Sasson et al., 2015). Although specific applications of blockchain, particularly cryptocurrency, have been criticized for their energy intensity to perform complex, computational processes (see Chapter 3), this Box discusses the use of distributed ledgers for decentralized data storage, which have lower energy consumption requirements (Platt et al., 2021).
>
> Distributed ledger technologies decentralize data ownership and governance by distributing data ownership across a network of "nodes" that each holds a copy of the entire information ledger, allowing individual actors or agents to contribute information and create consensus on which information should be added. For greenhouse gas data, each actor could integrate their accounting system as a node to contribute their data, while a distributed ledger would allow sharing of these data in a privacy-preserving and simultaneously transparent and trusted way (Schletz et al., 2022). Verifiable credentials are another digital technology discussed as a way to facilitate credible and trusted data sharing within this type of decentralized digital architecture. These credentials function similarly to a physical credential, such as a passport or driver's license, in the sense that they are proof that the holder owns a digital credential with certain verified characteristics (Sporny et al., 2022). To receive a verified credential, the data owner can submit a data derivative to a verifier that reviews and verifies the data and then, after confirming correctness, issues a verifiable credential for the respective data (Lux et al., 2020). The combination of verifiable credentials and zero-knowledge proofs thus helps manage anonymity, auditability, correlation across contexts, privacy, revocability, and traceability (Hyperledger, 2021; Sporny et al., 2022).
>
> Despite the nascence of data governance applications based on distributed ledger technologies, there is a growing number of case examples. Open Climate aims to be an integrated accounting platform with a particular focus on subnational and non-state actor climate data using distributed ledgers and other emerging technologies to develop a "nested accounting" approach. It strives to combine data across governance levels—from local to international—for the automatic inclusion of data that are needed to address present reporting delays and information asymmetries, as well as to support the United Nations Framework Convention on Climate Change Global Stocktake (Wainstein, 2019). Another example is the JLINX project, which provides a concept for distributed ledger-based decentralized data governance "to give both people and organizations control over data sharing and decision making (Fournier et al., 2022). Each party controls their own agent node which writes to their own micro-ledgers." A micro-ledger in this case acts like an individual blockchain to which only the owner has access and can exclusively determine data-sharing arrangements. These micro-ledgers provide data repositories that can be linked to verifiable credentials and thus provide an audit trail for interoperability and data provenance.

More research and development, particularly in exploring the ways artificial intelligence and ML approaches could improve GHG estimation methods, are critical to expanding their application.

Current Status and Uncertainties of Greenhouse Gas Emissions Estimates

Accuracy and uncertainty in GHG emissions estimates vary by type of emission-producing activity, spatial and temporal scale, GHG considered, and the technique used. In addition to the absolute uncertainty, uncertainty associated with a change in emissions over time is also important, particularly for emission reduction targets that are based on trends. Methods of assessing accuracy and uncertainty include examination of underlying data, emission factors, and approaches (Andres et al., 2014); comparisons of different activity-based emissions estimate products (e.g., Andrew, 2020; Friedlingstein et al., 2022); the inclusion of uncertainty estimates in the calculation for a product (e.g., Asefi-Najafabady et al., 2014; Solazzo et al., 2021); and independent assessment. In this section, we provide a summary of the current state of knowledge for emissions and indicate where activity-based methods are used to quantify uncertainties and in which areas atmospheric-based approaches also contribute.

The most accurate and precise estimates are available for emission-producing activities that have explicit economic value (e.g., energy production). National CO_2 emissions from fossil fuel burning are typically well characterized by economic data on fossil fuel trade and production, though there are large uncertainties for many countries. Thus, globally, fossil fuel CO_2 emissions are thought to be known with less than ±8 percent 2 sigma uncertainty (Andres et al., 2014) based on an intercomparison of different activity-based products, although many of these products do share input data such as IEA fuel statistics. While uncertainty in national fossil fuel CO_2 emissions totals for some nations with excellent data systems may be less than ±5 percent, countries with less well-developed energy data have been uncertainties of ±10 percent or more (Andres et al. 2014; Friedlingstein et al, 2022a; Marland, 2008). Smaller spatial and temporal scales (i.e., subnational, sub-annual) may increase uncertainty because approximation and/or estimation (e.g., proxy data) are often used that may be less directly linked with emissions and vary more across estimation techniques. However, in instances where direct estimation of emissions at smaller scales (e.g., at the point of combustion) is deployed, uncertainty may be lower than when aggregate amounts are downscaled. At 1–100 km resolution (i.e., relevant to cities and U.S. counties) the differences in fossil fuel CO_2 emissions among different products can be substantial and estimates vary depending on the approaches (Andres et al., 2016; Fischer et al., 2017; Oda et al., 2019).

Estimates of CO_2 emissions from land use and land-use change typically have much larger fractional uncertainties than fossil fuel CO_2. Globally, CO_2 emissions from land use and land-use change currently have uncertainties of roughly ±75 percent (±0.7 GtC yr^{-1} in 2020) (Friedlingstein et al., 2022). This estimated uncertainty has undergone large revisions recently using atmospheric-based approaches and may change in the future as different methodologies are compared in more detail (Bastos et al., 2021; Grassi et al., 2021; Petrescu et al., 2020). For individual countries or regions, uncertainties in CO_2 emissions from land use and land-use change can be more than ±100 percent (McGlynn et al., 2022).

Global total methane emissions have been estimated to ±5 percent by atmospheric-based approaches using measurements of the atmospheric methane growth rate and knowledge of the methane lifetime; however, highly uncertain natural emissions from wetlands, reservoirs, and other sources limit the atmospheric-based constraints on global anthropogenic emissions (Figure 2-7) (Saunois et al., 2020). Using isotopic measurements to isolate fossil fuel-derived methane emissions suggests global emissions of methane from the fossil fuel industry are 50 percent larger than inventory estimates (Hmiel et al., 2020; Schwietzke et al., 2016). The uncertainty of methane emissions from anthropogenic sources was documented in a 2018 report from the National Academies of Sciences, Engineering, and Medicine (NASEM, 2018). Other evidence of underestimated methane emissions from fossil fuel sources comes from atmospheric-based studies showing strong natural gas emissions in urban areas (Saboya et al., 2022; Sargent et al., 2021), areas of gas production (Alvarez et al., 2018), and the presence of super-emitters (Lauvaux et al., 2022) (see Case Study 4.4). Anthropogenic methane emissions from agriculture and waste also have large uncertainties (up to ±70%) (Solazzo et al., 2021) although there have been fewer independent studies to assess their accuracy (e.g., Miller et al., 2019).

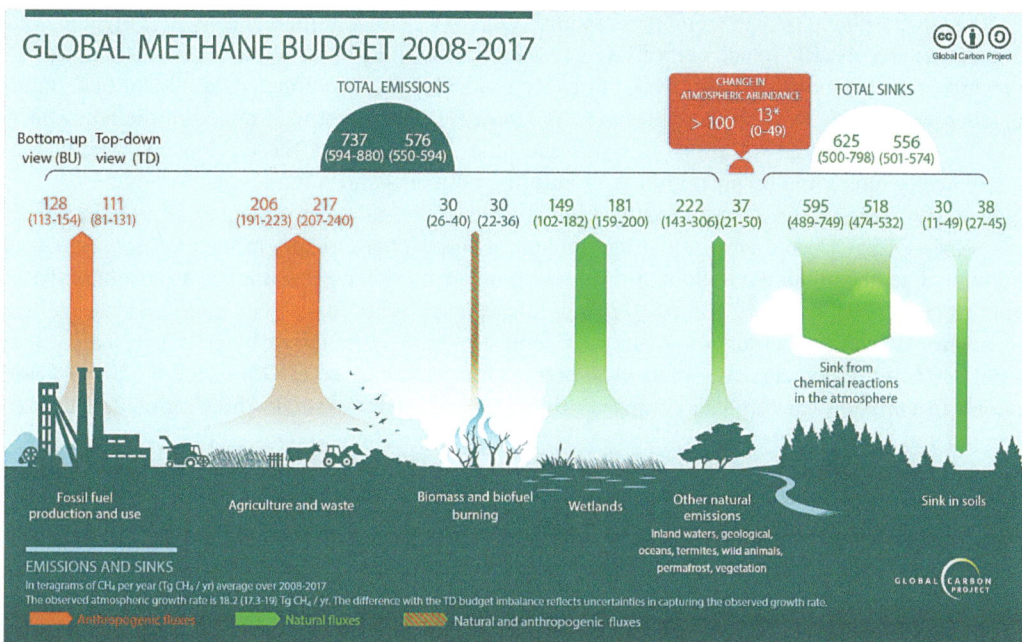

FIGURE 2-7 Global methane budget from 2008–2017 in teragrams of methane per year. The ranges show ±1 standard deviation. SOURCE: Global Carbon Project.

For N_2O, the largest emission sources are from agriculture and they are known to be highly uncertain (up to ±200%) because they result from the processing of nitrogen fertilizers in soil that can depend on environmental conditions (Solazzo et al., 2021). Atmospheric measurements indicate N_2O emissions are underestimated globally (Thompson et al., 2019), although good correspondence has been found between atmospheric data and inventory estimates for some countries (Ganesan et al., 2015).

For the industrially-produced fluorinated gases (i.e., CFCs [chlorofluorocarbons], HCFCs [hydrochlorofluorocarbons], HFCs, PFCs [perfluorocarbons], SF_6 [sulfur hexafluoride], and NF_3 [nitrogen trifluoride]), their global emissions can be well constrained by atmospheric-based approaches using measurements of the atmospheric growth rate and knowledge of the lifetime for each gas since there are no natural emissions. Atmospheric measurements have revealed unexpected emissions of several fluorinated gases, including CFC-11 (Box 2-7) (Montzka et al., 2018; Rigby et al., 2019), HFC-23 (Stanley et al., 2020), and HCFCs (Vollmer et al., 2021). For the HFCs, most existing inventory estimates do not report uncertainties (Flerlage et al., 2021). Moreover, not all emissions are reported because many countries are non-Annex I Parties to the UNFCCC that are not required to report HFC emissions, although most are now required to report production and consumption (not emissions) as part of the Kigali Amendment to the Montreal Protocol. Emissions from "banks" (stocks of CFCs, HCFCs, and HFCs in existing equipment and foams) are poorly constrained and contribute to emissions uncertainty (UNEP, 2019, 2021b; WMO, 2021). Uncertainties in regional fluorinated gas emissions from atmospheric based studies can range between ±20 percent to ±80 percent, where both underestimates and overestimates have been found for fluorinated gas emissions from individual countries (Flerlage et al., 2021; Manning, 2011).

For tropospheric ozone, GHG precursors (CO, NMVOCs [non-methane volatile organic compounds]), aerosols (BC [black carbon], OC [organic carbon], PM [particulate matter]), and aerosol precursors (SO_2 [sulfur dioxide], NO_x [nitrogen oxides], NH_3 [ammonia]), the global uncertainty is much greater due to their short lifetime. As a result, they are not well mixed in the atmosphere and are subject to large variations in space and time. Since most of these are air pollutants, they are typically monitored by air pollution monitoring networks and satellites.

Tropospheric ozone is not directly emitted, rather emissions of precursors including methane, NMVOCs, CO, and NO_x affect the amount of tropospheric ozone in the atmosphere (ozone burden). The uncertainties in global tropospheric ozone burden assessed using atmospheric-based approaches are on the order of 5 to 10 percent (Szopa et al., 2021).

Activity-based uncertainties of anthropogenic emissions of black carbon (BC) typically range from −30 to +120 percent and −40 to +130 percent for organic carbon (Klimont et al., 2017). Large ranges in emissions factors can contribute to higher uncertainties, as in the case of flaring (e.g., Conrad and Johnson, 2017).

Improved accounting of these short-lived pollutants has direct implications for understanding current and future climate impacts in addition to impacts on human health and food productivity. For example, incorrect anthropogenic aerosol emissions over China resulted in flawed model projections (Wang et al., 2021) with important implications for the net anthropogenic aerosol forcing trend (Quaas et al., 2022). Similarly, accounting only for CO_2 emissions from aviation without accounting for non-CO_2 emissions and contrails could leave significant warming unaddressed and compromise meeting temperature targets (Brasseur and Gupta, 2010; Brazzola et al., 2022; Wuebbles et al., 2007).

BOX 2-7
CFC-11: Detection and Resolution of Unexpected Emissions of a Banned Substance

Atmospheric measurements of the concentrations of chlorofluorocarbon (CFC) gases have shown a peak and then a decline since the adoption of the Montreal Protocol on Substances that Deplete the Ozone Layer (Montreal Protocol). This international treaty phased out the production and consumption of ozone-depleting substances including CFCs. Under the Montreal Protocol, new production of CFCs was banned after 2010; however, in 2018 atmospheric scientists reported that the decreasing trend in CFC-11 had unexpectedly slowed (Montzka et al., 2018). The slowing of the trend could only be explained by new production of CFC-11, in violation of the Montreal Protocol. Initial analyses were refined to identify the primary source of the emissions as Eastern China, although emissions may also have increased elsewhere (Montzka et al., 2021; Rigby et al., 2019). Scientists involved in the discovery of the unexpected emissions effectively communicated their findings to the Montreal Protocol organization, industry experts, and the public. Within a year, atmospheric measurements showed that emissions of CFC-11 declined and were again consistent with those expected from residual CFC banks in foams and equipment (Montzka et al., 2021). This episode clearly demonstrates the importance of atmospheric measurements for assessing compliance with laws and regulations, including how deliberate communication of the findings to stakeholders can generate a timely and effective impact on emissions.

3

Structural and Technical Limitations of the Current Greenhouse Gas Emissions Information Landscape

The approaches for quantifying greenhouse gas (GHG) emissions information discussed in Chapter 2, particularly activity- and atmospheric-based approaches, have been widely utilized by the scientific and regulatory communities to support national and regional emissions reporting, data collection, and research advances. However, a range of challenges have inhibited the usefulness of emissions information to support decision making. These challenges relate to both the institutions/structures and the technical work of developing inventories and other information. This chapter provides an overview and specific examples of these challenges, motivating the development of a framework for the evaluation of emissions information introduced in Chapter 4, and recommendations for improving future GHG information development and governance in Chapter 5. In both this chapter and the following chapter, we primarily focus on the national-to-global scale, in line with the Statement of Task.

Structural Limitations of the United Nations Framework Convention on Climate Change National Reporting Process

Limitations to the structure of the United Nations Framework Convention on Climate Change (UNFCCC) reporting process have hampered the completeness of global and national emissions inventories. The UNFCCC introduced the "common but differentiated responsibilities" concept that led to different standards and reporting requirements for Annex I countries and non-Annex I countries that have more reporting flexibility (see Chapter 2, Figure 2-3). Though justifiable from an international negotiating perspective, these differences in reporting standards have led to incomplete aggregate data at the United Nations—for example, around 45 countries have not reported emissions data since 2009 (Eilperin and Mooney, 2021). In addition, UNFCCC reporting requirements only include anthropogenic GHG emissions and sinks across five sectors. By convention, emissions from international marine bunker fuel and aviation are not part of national inventories, but may be reported separately. For the agriculture, forestry, and other land use sector, it can be difficult to identify and attribute emissions to human causes because these areas can function as both GHG sources and sinks due to natural and anthropogenic processes, contributing to large uncertainties in data and measurements (Perugini et al., 2021). Only Annex I countries are currently required to estimate these uncertainties. Non-Annex I countries are not required to produce uncertainty estimates.

The Intergovernmental Panel on Climate Change (IPCC) guidelines have defined source categories for reporting national emissions designed to limit double counting or omissions. However, certain source categories have key emission sources reported under different categories. For example, the latest U.S. GHG inventory reports CO_2 emissions from all combustion sources within the natural gas value chain under the Fossil Fuel Combustion source category, while methane and CO_2 emissions from flaring are reported under the Natural Gas Systems source category. This makes it difficult to compare national GHG inventories with inventories developed by the regulated and scientific communities (see Table 2-1), especially for communities or decision makers unfamiliar with the IPCC reporting structures.

In general, the IPCC process has long grappled with the incorporation of peer-reviewed information and timeliness and has been limited by the complexity of the report writing process and sign-off by governments. The development process for the IPCC inventory guidelines also introduces issues that hinder the effectiveness of GHG inventory reporting. The expert synthesis process, by which the IPCC guidelines are updated, has remained unchanged since 1996. The four-stage process of expert written and revised drafts cannot always include peer-reviewed research due to publication deadlines, thereby hindering the incorporation of the latest scientific studies into future updates (Yona et al., 2020). It was not until 2015 that Annex I countries were required to use the 2006 guidelines as part of their inventory submittals (UNFCCC, 2014, Decision 24/CP.19). Similarly, the use of the 2019 refinements is optional for inventories submitted as part of the Enhanced Transparency Framework (ETF). While not required, the tiered architecture of the IPCC process does not prevent countries from using higher tiered (tier 2 and 3) approaches that reflect the latest literature and are relevant for the country.

Barriers to Capacity Building

GHG emissions inventory research, in general, has been limited in developing countries because of a lack of technical expertise and scientific capacity, often the result of limited financial resources and investments in institutions (Kim et al., 2022a). Moreover, climate change disproportionately affects developing countries (IPCC, 2006); thus, domestic funding often prioritizes adaptation to climate change rather than mitigation (Kim et al., 2022a; López-Ballesteros et al., 2018). While many local and regional efforts can be supported through capacity building, this section focuses on the UNFCCC reporting processes because it is a requirement for Parties and the foundation upon which many subnational emissions inventory efforts have been based.

Adopted at COP-21 (21st session of the Conference of the Parties), the Paris Agreement developed the ETF for action and support "in order to build mutual trust and confidence and to promote effective implementation" (UNFCCC, 2015, Article 13.1). The purpose of the ETF is to improve the understanding of climate mitigation approaches and track the progress countries are making toward their nationally determined contributions (Article 13.5). The ETF strengthens and improves existing transparency arrangements, including national communications and biennial update reports (BURs) while considering the different capacities and circumstances of least developed countries and small island developing states (Articles 13.3 and 13.4). Under this framework, each Party agrees to provide a regular national inventory report, prepared by using IPCC's best practices (Article 13.7(a)).

Financial, technical, and capacity-building support is provided to assist with UNFCCC reporting requirements (UNFCCC, 2015, Articles 9, 10, 11). The Global Environment Facility (GEF) provides financial support and the Consultative Group of Experts provides technical support.

Still, barriers to capacity building remain. GEF funds are contractually provided, meaning they are provided to pay for a project (e.g., to prepare a BUR). Although GEF has fulfilled its function to fund the preparation of reporting projects, there can be a gap in funding cycles and, without domestic or other sustained funding, developing countries often lose the required technical knowledge—including staff, data, and methods documentation—to support GHG reporting. Subsequently, many countries must re-acquire all the necessary components for each new GHG inventory report (Damassa and Elsayed, 2013). Similarly, the lack of institutional arrangements (e.g., specific agencies responsible for reporting) within a country hinders the ability to sustain GHG reporting (Breidenich, 2011).

Kawanishi and Fujikura (2018) evaluated the capacity of non-Annex I countries to sustain national GHG inventory reports (i.e., NCs and/or BURs). They used criteria developed by Damassa and Elsayed (2013) and added two separate criteria related to the issue of sustained funding: one criterion for "external funding" and another for "domestic funding." Kawanishi and Fujikura (2018) found a significant difference in all eight criteria except for "external funding" between countries that have conducted more regular reporting and countries with limited to no reporting, indicating external funding has played a limited role in capacity building. The most significant difference between the two groups was in the "institutional arrangements" criterion. Criteria "domestic funding," "archiving," and "data sharing agreements" were the next most significant (Kawanishi and Fujikura, 2018). Ultimately, Kawanishi and Fujikura (2018) found that access to GEF financial support did not necessarily translate to enhanced capacities to facilitate regular inventory reporting.

In addition to financial limitations, barriers to capacity can be a result of political, institutional, and/or technical factors that hinder or prevent the preparation of regular GHG inventory reports (Damassa and Elsayed, 2013; IPCC, 2006). In an assessment of 37 Asian developing countries, basic technical capacity (e.g., statistics and scientific expertise) in each country was crucial to support the improvement in overall GHG inventory reporting capacity (Umemiya et al., 2017). In a discussion among stakeholders, they noted important issues related to data and metadata and highlighted the importance of data and knowledge sharing (e.g., methodological guidance and research results) (López-Ballesteros et al., 2018). It is important to consider country- and context-specific needs (Kawanishi and Fujikura, 2018) and to ensure needs are provided and prioritized by developing countries themselves (Kim et al., 2022b).

Current Institutions and Their Limitations

As discussed in Chapter 2, there is a wide range of institutions involved in generating, collecting, curating, and disseminating GHG emissions information at global, national, and subnational scales. Figure 3-1 provides some examples of existing organizations, programs, and projects across spatial scales.

At the global scale, this includes the Committee on Earth Observation Satellites coordinating satellite observation systems; the World Meteorological Organization program Integrated Global Greenhouse Gas Information System (IG^3IS) seeking to extend the observational capacity for GHGs to all scales and provide information to users from national governments, to cities, to facilities; the International Methane Emissions Observatory (IMEO) (see Box 3-1) focusing on methane emissions observations; other Earth observation portals such as Copernicus and National Aeronautics and Space Administration; and a number of institutions devoted to developing global inventories such as the Carbon Dioxide Information Analysis Center and the International Energy Agency (see Table 2-1 for a list of examples). The UNFCCC and IPCC have a unique role given their linkage to the UNFCCC national reporting requirements, with a separate set of issues described earlier in this chapter.

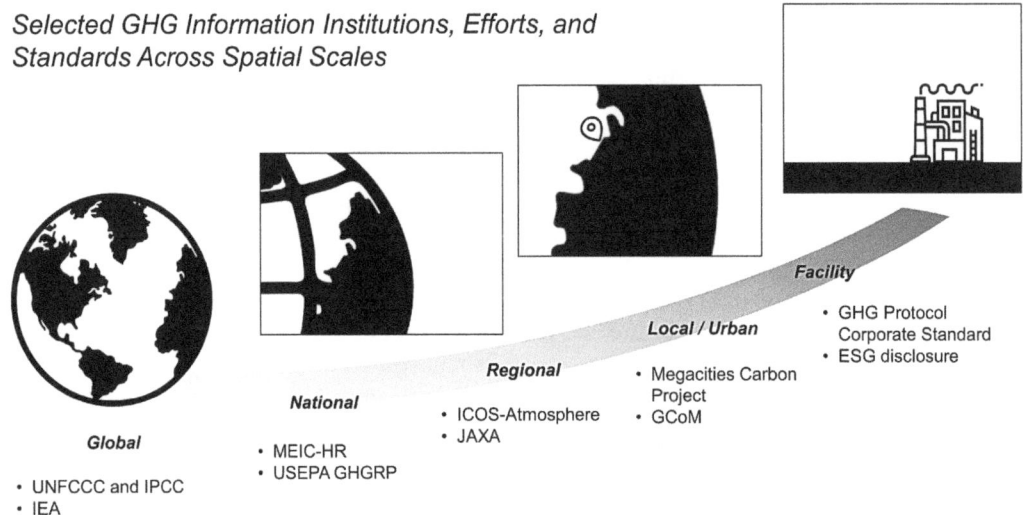

FIGURE 3-1 Selected greenhouse gas (GHG) emissions information inventories, institutions, and standards from global to facility spatial scales mentioned in the report. Spatial scales are shown on a continuum and the examples provided here are meant to be demonstrative rather than comprehensive. NOTES: ESG, environmental, social, and governance; ICOS, Integrated Carbon Observation System; IEA, International Energy Agency; IPCC, Intergovernmental Panel on Climate Change; GCoM, Global Covenant of Mayors; JAXA, Japan Aerospace Exploration Agency; MEIC-HR, Multiresolution Emission Inventory for China – High Resolution; UNFCCC, United Nations Framework Convention on Climate Change; USEPA GHGRP, U.S. Environmental Protection Agency GHG Reporting Program; WMO IG^{3}IS, World Meteorological Organization Integrated Global Greenhouse Gas Information System.

BOX 3-1
International Methane Emissions Observatory (IMEO)

The International Methane Emissions Observatory (IMEO) was launched at the 2021 G20 (Group of Twenty) Rome summit with the goal of producing an empirically verified global dataset of methane emissions across all methane emission sectors. The IMEO will review and support scientific studies across multiple scales and technologies to integrate the relevant information into a coherent and actionable dataset for stakeholders, especially to support the commitments made under the Global Methane Pledge. The IMEO's initial focus will be on the fossil fuel (oil and gas and coal mining) sectors, with plans to expand to other anthropogenic sources such as agriculture and waste. The IMEO holds the potential to serve as the global hub for anthropogenic methane emissions data and an example of a purpose-built hybrid approach.

At the national scale, most countries have national institutions responsible for collecting climate change statistics according to the UN Statistical Commission Global Consultation on the Global Set of Climate Change Statistics and Indicators (UNSD, 2022), but the responsible institution or institutions vary. Most of the national statistical organizations have not yet developed specialized climate change surveys, reflecting that climate-specific indicators are a relatively new area for standardized tracking and a lack of mandates and capacity for collecting this information. The consultations also found that a minority of developing (non-Annex I) countries had requested technical assistance from regional and international organizations. Exceptions to this are worth noting, such as the U.S. Environmental Protection Agency, which archives an increasingly rich complement of GHG emissions information both as part of the UNFCCC process and for domestic regulatory purposes.

As with research activities taking place at the global scale, there are also individual research efforts at the national scale that generate increasingly complex and granular GHG emissions information. However, these efforts typically lack the institutional infrastructure needed to systematically serve GHG emissions information to the wide spectrum of decision makers over long periods of time.

Drawbacks of the current global institutional structures include the lack of a framework for users to easily navigate and find relevant and useful information. The current landscape requires users to know what resources exist, how to find them, visit each resource to see what information is available, and then interpret the quality and utility of that information from often technical and hard-to-access documentation. Furthermore, the process by which new scientific research is integrated into analytical frameworks to develop inventories is opaque. These challenges are not unique to GHG emissions information. The air quality and meteorological communities face these same challenges, and the strategies, lessons learned, and infrastructure developments provide a template and collaborative opportunity for GHG-focused efforts. Furthermore, the synergy of GHG, air quality, and meteorology can be leveraged to advance science and better utilize resources, including analytical and measurement capabilities.

Technical Challenges with Current Approaches

Limitations to Activity-Based Approaches

As described in Chapter 2, the most general representation of the method at the core of most activity-based approaches involves multiplying an emission factor with activity data for the time period the inventory is developed. The quality of emissions inventories developed using this approach is thus often determined by the quality of the emission factor and activity data inputs.

Activity data. Activity data can represent a wide variety of information such as direct flux observations, models, survey data, remotely-sensed data, and socioeconomic indicators. We do not exhaustively review all forms of activity data here but use a series of example activity-based information that provides common example limitations to activity data.

The simplest example involves the estimation of energy-related GHG emissions from fossil fuel combustion. Although other techniques exist (e.g., ratios to monitored local air pollution), country-scale energy statistics are the most common activity data used to estimate GHG emissions from energy-related activities (IEA, 2021c). As described in Chapter 2, energy statistics are acquired through national questionnaires or surveys (Andres et al., 2012) but they often lack transparency. This lack of transparency presents a potential limitation to the reliability and accuracy of this foundational energy information and is a key source of uncertainty for global inventories that rely on these activity data (e.g., Hoesly and Smith, 2018; Marland, 2008). The major energy statistics sources often provide national energy or fuel use datasets that are updated and revised on an annual basis. While the national energy or fuel use data do not provide enough information for emissions to be calculated beyond country and annual scales, the consistent and complete time series of national data should ensure robustness of interannual emissions. Thus, the resulting emissions can be used to support the evaluation of emission changes from a base year to target emission reductions (Oda et al., 2021) and the uncertainty discussed above may not be an issue for certain applications, such as UNFCCC reporting, that use these emissions estimates. Uncertainties associated with conventional fuel activity data at national scales are considered to be small, especially for CO_2, due to the robust definition of the system boundary (i.e., apparent consumption approach) and the direct connection of fuel and emissions (i.e., no chemistry other than oxidation) (Oda et al., 2021).

Additional energy information may be available at finer spatial scales, such as estimates from energy utilities, fuel accounts from individual energy industry companies, and fuel sales at the retail level. However, this information is often not universally available or archived systematically. In addition, defining the system boundaries of these activity data is more challenging than at the national level, particularly for certain sectors such as transportation (Oda et al., 2021). Beyond energy statistics, there is a lack of coordinated collection systems for activity data required to generate activity-based emission estimates. Unlike at the national level, there is no systematic evaluation of activity data at finer spatial scales to detect potential errors and biases.

Beyond simple fuel statistics, other forms of activity data pose unique challenges. For example, direct flux monitoring, such as those represented by continuous emission monitoring (CEM) devices present in large effluent stacks, is a promising approach to unambiguously quantify GHG emissions from an emitting activity (Gurney et al., 2016; Quick, 2014). However, CEMs have mostly been applied to power production facilities in high-income countries and require ongoing independent calibration and maintenance. Furthermore, where there are no regulatory requirements or market drivers, large effluent stack emitters are typically private actors and the expansion of CEMs must overcome proprietary and privacy barriers for wider deployment.

Given the importance of onroad vehicular GHG emissions, a common approach to estimating GHG emissions independent of fuel statistics is via measurements of the vehicle miles or kilometers traveled (Gately et al., 2017). Though a useful independent estimate of onroad emissions, these measurements require dense traffic monitoring across a statistically representative sample of road segments, often prohibitive to countries with a large geographic extent and/or limited resources. However, new forms of nontraditional activity data, such as ubiquitous cell phone and GPS locational information, offer the potential to overcome this limitation (Gately et al., 2017).

Estimating managed vegetative carbon uptake and emissions often requires the estimation of vegetation metrics such as vegetation cover or mass, growth rates, or soil composition (Oertel et al., 2016; Sha et al., 2022). These metrics can be estimated from the extrapolation of intensive plot studies or estimated from satellite-based measures of activity such as the Normalized Difference Vegetation Index or Solar Induced Fluorescence (SiF) (Kováč et al., 2022; MacBean et al., 2018). Extrapolation faces challenges of heterogeneity and sample representativeness particularly because the managed vegetation is considered regionally specific. Satellite-based approaches face challenges associated with cloud cover, overpass limits, and biome representativeness.

Some of the limitations of activity-based approaches are linked to the specific GHG considered. For example, methane emissions are challenging to quantify because large portions of methane emissions are from unintentional or unaccounted release (in the form of natural gas) from extraction to end-use delivery for which there are limited to no representative statistics. This and other forms of unintended emissions create particular challenges (e.g., Chen et al., 2022a; Cusworth et al., 2020; see Box 3-2).

Many forms of activity data are generated through survey collection, which can take several years and result in emission estimates that have a lag of 1 or 2 years (NRC, 2010). The implication of this time lag is that stakeholders do not have dynamic or contemporaneous estimates of emissions to better inform decisions, particularly at subnational levels (see Box 3-2). However, it is important to note that national emissions inventories are complete emission time series based on updated activity data over the period of interest. While there are many activity-based emissions estimates from the research community, they typically do not update their emissions estimates to reflect updated activity data. Emissions information products that are based on outdated activity data are challenging to compare to other estimates. Mostly because of the COVID pandemic, innovations

> **BOX 3-2**
> **Capturing Unknown Emissions**
>
> The most common activity-based approach for global/national energy-related CO_2 emissions is based on fuel statistical data. For example, global fossil fuel CO_2 emissions are based on the production of fuels, and country emissions are based on apparent fuel consumption (Marland and Rotty, 1984). While this approach often yields robust estimates at those scales on an annual basis, these approaches are inherently limited in their ability to account for unreported emissions (e.g., equipment failure and leakage) or military and wartime activities. For example, peacetime military activities have been estimated to contribute approximately 6 percent of global greenhouse gas emissions (Belcher et al., 2020; Bun, 2022; Parkinson, 2020; Pereira et al., 2022). Under the current reporting framework, the reporting of fuel consumption for multilateral operations (i.e., emissions from fuels used in multilateral operations according to the Charter of the United Nations, including emissions from fuel delivered to the military in the country and delivered to the military of other countries) is voluntary and thus may be poorly captured. In addition, non-peacetime emissions can be poorly characterized due to damage to data collection systems (Bun, 2022). Wartime military emissions can be significant with one study estimating that burning oil and gas in Kuwait during the 1991 Gulf War emitted 2–3 percent of global anthropogenic CO_2 from fossil fuel use and 10 percent of particulate matter from global anthropogenic fossil fuel burning (Boden et al., 2011; Linden et al., 2004). The actual locations of emissions are also thought to be uncertain as access to information on military deployment is limited (Olsen et al., 2013).

in inventory work have begun to address this limitation, such as with studies generating global estimates in near real-time (Liu et al., 2020b). However, these near-real-time estimates may have errors due to the use of activity data that have not been thoroughly evaluated (Oda et al., 2021).

Finally, recent years have seen the use of "nontraditional" activity data—for example, facility-level operational data (e.g., Liu et al., 2020a), cell phone data, and human mobility data (e.g., Forster et al., 2020). These data are typically used to produce higher time and spatial resolution emission estimates beyond annual country scales. However, while these nontraditional activity data are potentially useful, it can be challenging to ensure the spatial and temporal representativeness of the data for emissions calculations (Oda et al., 2021). The importance of a mechanistic understanding of high-emitting activities has been highlighted in multiple studies (NASEM, 2018; Schwietzke et al., 2017).

Very high-resolution optical remote sensing, available primarily from commercial satellites, is being explored to characterize emitting infrastructure, particularly because it is both more granular than the current global inventories and offers global data availability (Jervis et al., 2021; Jongaramrungruang et al., 2019). However, challenges remain in the interpretation and classification of imagery and the process by which GHG emissions are estimated (i.e., through emission factors or other transfer functions) (Jervis et al., 2021; Varon et al., 2018).

Emission factors. Given the wide variety of activity data, emission factors or transfer functions from the activity to the estimate of emissions are equally diverse. This section will similarly avoid an exhaustive treatment of the many emission factor estimation techniques, focusing instead on a few representative examples with a description of their common limitations.

As with the activity data, the simplest and most widely used emission factor data, particularly when considering fuel combustion-related GHG emissions, is the carbon content of fuel. It is critical for these emission factors to be representative of the country, area, sector, and time of interest to provide accurate combustion-related emissions estimates (IPCC, 2006, 2019a). Emission factors can be found in the scientific literature or a database if not otherwise available. While emission factors in the scientific literature or databases (e.g., IPCC Emission Factor Database[1]) often include references for traceability and transparency, the level of representativeness varies over countries, regions, and sectors. For example, emission factors for fossil fuel combustion emissions are relatively well known; however, emission factors for non-fossil fuel burning emissions (e.g., biomass burning) can be highly uncertain due to the heterogeneity of the fuel burned and the combustion conditions (e.g., Akagi et al., 2014). Furthermore, carbon content is often measured in physical units (e.g., mass carbon per gallon of fuel), which requires additional information about the heat content of fuel and imposes additional demands for observations and uncertainty quantification.

Combustion or fossil fuel-related emission factor estimations have been found to not align with peer-reviewed scientific studies and actual measured emissions in various sectors, creating a GHG accounting gap between GHG inventories and observations (Yona et al., 2022). For example, multiple studies in the oil and gas sector (e.g., Alvarez et al., 2018; Chan et al., 2020; Chen et al., 2022b), including longer-term measurements, have revealed that facility-level methane emissions are higher than those reported using conventional inventory methods (Wang et al., 2022a). This is mainly due to the continued use of outdated emission factors or poorly characterized and unaccounted emission sources (Cusworth et al., 2021b; Gordon, 2022; NASEM, 2018; Omara et al., 2022; see Case Study 4.4).

When considering onroad GHG emission estimation approaches that do not use bulk fuel statistics, several transformations are required. For example, the fuel efficiency of vehicles specific to country, region, vehicle class, driving condition, temperature, and maintenance level is critical to the accurate estimation of onroad emissions along with the fleet vehicle class distribution.

Transformation of vegetation characteristics to net emissions or uptake often requires techniques that span a wide spectrum of complexity from allometric relationships to complex carbon cycle plant or soil models. Both photosynthetic uptake and respiration emissions can be modeled and integrated (Velasco et al., 2013, 2016; Weissert et al., 2017).

Where relevant, emission factor databases that are widely used are often out of date and not regularly updated based on new information. For example, the IPCC Emission Factor Database, developed in 2006, has emission factors created using data from the 1990s based on Organisation for Economic Co-operation and Development countries (Chen and Brauch, 2021), and the same IPCC emission factors are used today.

Interest in consumption-based accounting perspectives faces numerous challenges. For example, translation of commodity flows at the national scale must overcome approximations related to the use of global average emission factors, coarse commodity categorization, and data transparency barriers (Iacovidou et al., 2017; Peters, 2008).

Furthermore, corporate inventories developed annually as part of either regulatory or voluntary requirements, such as Corporate Sustainability Reports, generally employ a similar accounting framework on the use of emission factors and activity data to report their direct (Scope 1) and indirect (Scope 2 and 3) emissions and have similar limitations as national inventories. The emission factors employed in corporate inventories are generally based on national or regional-level studies and, as such, may not be representative of a facility at the corporate level or across supply chains (Roman-White et al., 2021; Stern, 2022).

Spatial challenges. Activity-based inventories include varied levels of spatial information. Emissions estimates can be disaggregated using geolocation information of sources and/or proxy (surrogate) data to estimate subsector emissions. Spatially explicit emissions can be calculated at the location of sources and used to provide subnational (province/state, cities, and facility levels) emissions, though uncertainties associated with spatial estimates can be large.

---

[1] https://www.ipcc-nggip.iges.or.jp/EFDB/main.php

Activity-based inventory approaches are also applicable for subnational scales, such as state/province, cities, and individual facilities. Approaches used at subnational scales are often based on the same approaches used at national scales (e.g., community protocol for cities[2]), and thus are comparable to estimates from other subnational scales; however, emissions estimates may not be comparable because of the choice of emission factors and activity data. Particularly for activity data, the system boundary of subnational entities such as cities can be difficult to define (e.g., traffic emissions) and may not be the same as at the national level (Scope 3). For example, by disaggregating spatial data into $0.1° \times 0.1°$ monthly gridded inventories of methane emissions for the United States, Maasakkers et al. (2016) were able to illustrate limitations and large errors in the spatial allocation in other inventory methods that can impact inverse analysis, if used as an a priori.

Examples of high-resolution GHG inventories have been developed for Europe, the United States, Poland, and Ukraine (Bun et al., 2007, 2019; Gately and Hutyra, 2017; Gurney et al., 2009, 2020b; Ivanova et al., 2017; Moran et al., 2018, 2022). In the case of Bun et al. (2019), emissions were disaggregated from province to source scales and categorized into IPCC sectors for CO_2, methane, N_2O, some F gases, and air pollutants. The emissions estimates were combined with corresponding geospatial source information (e.g., point, line, and areas) and provided as sectoral vector digital emission maps. Depending on source types, additional data, such as fuel use and type, were utilized to spatially downscale emissions by considering the nature of emissions sources. Unlike gridded products, this approach makes it possible to implement spatial emission analysis at key subnational administrative scales by correctly addressing administrative boundaries (Figure 3-2) and providing spatially resolved emission maps (up to 100 m). The efforts in Ukraine and Poland have done considerable work on quantifying the uncertainty associated with sectoral emissions estimates, spatial estimates, and these downscale/allocation procedures (Bun et al., 2007, 2019).

Examples in the United States avoid much downscaling with data at the asset scale such as facility-scale emissions stack monitoring and onroad vehicle activity at the individual road segment

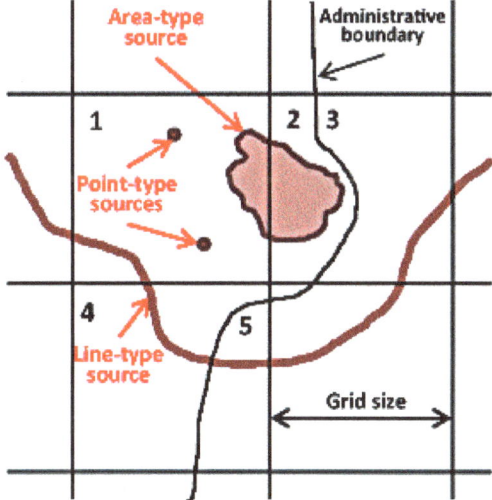

FIGURE 3-2 Example of combining diverse emissions sources into gridded and administrative boundaries in the development of activity-based emissions inventories. SOURCE: Bun et al., 2019.

---

[2] https://ghgprotocol.org/greenhouse-gas-protocol-accounting-reporting-standard-cities

scale (Gately and Hutyra, 2017; Gurney et al., 2009, 2020a). Further research at the urban scale has resulted in emissions quantification at the individual building and road segment scale (Gurney et al., 2012, 2020b). However, even in data-rich countries such as the United States, these more spatially-resolved techniques face challenges. Independent evaluation is limited and atmospheric-based approaches must be constructed at finer scales to provide an alternative estimation. Furthermore, many of the data elements used in more spatially-resolved approaches are self-reported, raising challenges of validation, accuracy, representativeness, and trust.

The limitations of activity-based approaches noted above hinder timely action on mitigation strategies, especially in the case of short-lived climate forcers such as methane, for which there are significant uncertainties in existing emissions inventories.

Limitations of Atmospheric-Based Approaches

Atmospheric-based approaches utilize atmospheric measurements and models to quantify GHG emissions. However, atmospheric measurements can be limited in terms of precision, continuity, and coverage (see Box 3-3). In addition, models of atmospheric transport and chemical processes

---

BOX 3-3
CO_2 Emission Changes During COVID-19 Lockdowns

The COVID-19 pandemic caused a series of lockdowns beginning in 2020 to restrict the spread of the virus. Economic and human activities including industry and transportation were curtailed in many countries, although at different times and to different extents. Improvements in air quality were visible to the public in various locations. However, changes in air quality were more complicated, with ozone pollution increasing in many cities and some air quality changes masked by weather (e.g., Deroubaix et al., 2021; Grange et al., 2021; Siciliano et al., 2020; Zhao et al., 2020). Concurrent with changes in air pollutant emissions, analogous reductions in fossil fuel CO_2 were also speculated about. This episode provided an unprecedented opportunity to test how different greenhouse gas (GHG) emissions information systems could detect unexpected GHG emissions changes, while also highlighting current challenges of providing timely, responsive GHG emissions information to decision makers and the public. It kick-started the development of novel approaches to estimate emissions, speed up the generation and communication of GHG emissions information using near-real-time data, and better integrated various datasets to produce GHG emissions information.

For activity-based approaches, higher temporal resolution and lower latency than is typically available were needed to assess the changes in GHG emissions from lockdowns. Activity data such as electricity generation/use, traffic data, mobile phone and internet search data, and aircraft departures were quickly combined with more traditional data and emissions estimates to produce the first estimates of the lockdown effect on the 2020 annual global emissions (–6% compared to 2019) in April 2020 (Le Quéré et al., 2020). Efforts to synthesize near-real-time data of GHG emissions have accelerated since 2020 (Liu et al., 2020b). Challenges remain in near-real-time estimation approaches due to the limited data availability and geographic representativeness of the activity data for sectoral emissions changes (Gensheimer et al., 2021; Oda et al., 2021).

The latency of atmospheric-based approaches is also traditionally too long to quickly inform users about GHG emissions changes within days or months, although by May 2020, emissions changes observed by eddy-covariance measurements in Europe had already been reported (Papale et al., 2020). Changes in CO_2 emissions have since been reported retroactively for several urban regions using ground-based atmospheric data, including innovations such as novel fossil fuel CO_2 tracers (Pickers et al., 2022), low-cost sensor networks (Turner et al., 2020), and citizen science (Turnbull et al., 2022). Satellite measurements of NO_2 have been used to infer changes in regional fossil fuel CO_2 emissions (Zheng et al., 2020), while direct observations of reductions in CO_2 emissions by satellite CO_2 measurements were found to be difficult due to cloud-related data gaps, small signals relative to measurement uncertainty, and the confounding effects of the natural carbon cycle (Chevallier et al., 2020).

introduce uncertainties and assumptions, impacting the accuracy of an emissions estimate from an atmospheric-based approach.

Major challenges for atmospheric-based studies relate to the continuity and comparability of atmospheric measurements. To resolve small gradients in the atmosphere, measurements must be highly precise and traceable to a reference (Brewer et al., 2018). The expert community of GHG measurement scientists regularly updates a set of recommendations for measurement and calibration techniques (WMO/IAEA, 2020). However, the capacity of central calibration laboratories may not be sufficient for future expansion in measurements and new systems may be needed. Calibration of satellite data and verification of its compatibility with other data requires ground-based remote sensing and aircraft measurements, and best practices are under ongoing development. Continuity of measurements is also needed to establish reliable atmospheric-based GHG information that can be integrated into decision making, but often atmospheric measurements are tied to specific research projects or political budget cycles and do not have secure, long-term support.

Another key challenge for atmospheric-based studies of CO_2, methane, and N_2O is the separation of anthropogenic and natural emissions. The most reliable (and similarly atmospheric-based) techniques use measurements of tracers that are sensitive to anthropogenic sources, which are associated with fossil fuel sources, but also biogenic sources in some cases. However, the sparseness of tracer data presently limits the use of these approaches. In some cases, uncertainties in emission ratios, emissions, or dynamics of these tracers are also limitations. For CO_2, only a small number of inversions exist or are in development where fossil fuel emissions are determined for regional/national scales (Basu et al., 2020; Graven et al., 2018; Petrescu et al., 2021) and city scales (Turnbull et al. 2019). These studies used radiocarbon, nitrous oxides, and carbon monoxide as fossil fuel combustion tracers. For methane and N_2O, isotopic measurements and tracers such as ethane have been used on scales ranging from global to local to separate different sources.

Inversion methods are the most widely used atmospheric-based approaches, and their development and limitations are the focus of the remainder of this section. There is a large body of scientific literature that discusses atmospheric inversions and the results of atmospheric inversion for inferring emissions (Deng et al., 2022; Friedlingstein et al., 2022; Jacob et al., 2022; Janssens-Maenhout et al., 2020; NASEM, 2018). Key points on the strengths and limitations of these approaches are briefly summarized here, in the areas of the datasets, inversion tools, and spatial and sectoral information.

There have been advancements in the tools for inversion analysis, especially as computing power has improved, and weather/dynamics models improve fidelity and resolution (Agustí-Panareda et al., 2019). Despite this progress, the transport in atmospheric models can still be an important source of error (Basu et al., 2018; Crowell et al., 2019; Peylin et al., 2013; Schuh et al., 2019). Atmospheric inversions utilize meteorological data and/or models to make connections between atmospheric observations and emissions. Thus, the error and uncertainty associated with atmospheric transport directly impact the resulting inverse estimates. The impact of atmospheric transport errors has been characterized by comparing results from different transport models (e.g., Crowell et al., 2019; Peylin et al., 2013), though using the model ensemble spread may be an insufficient approach to fully capture the transport errors. For example, Schuh et al. (2019) showed differences in meridional and vertical transport simulated in transport models commonly used in large-scale inversions (e.g., Crowell et al., 2019). Complex meteorology can also pose a challenge to inversions. There are limited atmospheric boundary layer height (ABLH; the lowest layer of the atmosphere in which GHG emissions get diluted observations) and there are large uncertainties in atmospheric transport models that need such observations to improve the representation of the ABLH. In addition, atmospheric inversions typically employ "offline" models driven by meteorological fields taken from global climate models (GCMs). Analysis from parent GCMs could mitigate transport errors (e.g., Yu et al., 2018).

Although atmospheric-based methods often involve air pollutant measurements, there has historically been limited collaboration between the GHG and air quality communities. The air quality community has a long record of developing and utilizing detailed emissions inventories using the same methods described in Chapter 2, and many of these inventories are tailored to cities that also account for the majority of GHG emissions. Not only do many air pollutants have common sources with GHGs, but air pollutant inventories also contain many of the same activity-based information relevant to developing GHG emissions inventories. Emissions inventories and measurements of air pollutants may not be co-located with GHG measurements or performed by different entities, hindering the ability of these independent measurements to constrain atmospheric-based estimates of GHG emissions. Thus, there would be large co-benefits for climate and human health from integrating studies of GHGs and air pollutants, as well as integrating mitigation policies for GHGs and air pollutants.

There has been a large increase in the number of remote sensing satellites for CO_2 and methane since 2010 when National Research Council (2010) was written. The plans for future missions are described in Chapter 2 and Appendix B. While new satellite measurements and advances in inversion models have allowed progress in some areas, challenges remain related to data coverage, limitations of sensitivity, and difficulty in separating sectors. In some cases, the in situ and/or remote sensing data are not sufficient to characterize the flux or emissions of interest when the magnitude of the emissions is below the sensitivity of the analysis (Jacob et al., 2022). Inversion analysis is impacted by uncertainty on the fluxes of natural systems that make it hard to quantify anthropogenic emissions. Even then, the end user is often interested in emissions from a specific sector, not the integrated total that is sensed by atmospheric measurements. There are methods under development to add information on specific sector sources to inversion analysis, such as the use of isotopic tracers and multiple atmospheric gases (see Chapter 2). The inclusion of additional variables related to the activity of the natural system (such as carbonyl sulfide and SiF) can be used to try to separate the terms, but this is still an area of development. For regional and city scales and point sources, the background concentrations must be well specified, as they can have a large influence on inversion results. If there are multiple emissions sources located together in one area, it is unlikely that atmospheric inversions will be able to distinguish them. In addition, a single inverse estimate provides little information about the accuracy of the inverse estimate, so often, multiple inversions are needed to gain some insight into the consensus of the inversions and perhaps infer the accuracy of the inversion.

Other barriers exist related to the community of practitioners, communication of results, and data sharing. The application of inversion models is complicated, requires large computing resources, and utilizes large, complex computer codes. A small group of researchers performs these analyses, and the techniques and descriptions can even be hard for other scientists to understand. Results are usually shared via publication in the scientific literature using terminology aimed at other scientists. Findings and implications are rarely widely communicated or put in language more useful to policy makers. Although many measurement datasets are freely available (see Chapter 2), there are some datasets collected by private entities that are not publicly available or are available for purchase. This is a barrier to transparency: even if analysis results are published, the underlying data cannot be accessed for others to use in analysis.

Challenges in Reconciling Activity- and Atmospheric-Based Approaches

Atmospheric-based approaches can provide an emissions estimate that is independent of an activity-based estimate if the atmospheric-based approach does not use a prior estimate from an activity-based approach. Evaluations of activity-based estimates can often be challenging due to the lack of evaluation data, such as alternative emissions estimates (e.g., Andres et al., 2016; Oda et al., 2019). Thus, assessment of an activity-based inventory's uncertainties is often limited. In particular cases, atmospheric-based estimates may hold the potential to examine the accuracy of activity-based estimates and detect and quantify errors and biases, especially if the errors are associated with a particular approach (e.g., conceptualization error). However, comparisons of estimates from the two approaches can be challenging due to the mismatch of the spatial and temporal domain and coverage as well as the granularity of the information. Furthermore, separating emissions and removal contributions to atmospheric-based approaches can also make such comparisons challenging. Some programs, such as the Global Carbon Project, are reporting comparisons between estimates from activity-based and atmospheric-based methods (e.g., Friedlingstein et al., 2022).

The uncertainty associated with atmospheric-based approaches is often significantly higher than that of activity-based approaches (Cheewaphongphan et al., 2019; Vaughn et al., 2018), although the uncertainty in activity-based approaches is often not quantified or well understood. Reconciling activity- and atmospheric-based approaches is often challenging due to the difference in model conceptualization. Activity-based approaches often define the domain of interest as a system boundary (e.g., country). However, it is challenging to match the domain of atmospheric-based estimates with the proper boundary to carry out a fair comparison. Smaller spatial domains, such as cities, have a better chance of successfully reconciling activity- and atmospheric-based approaches, particularly if dense observation networks are utilized (e.g., Lauvaux et al., 2020). While atmospheric-based estimates are more accurate at smaller scales, facility-level emission estimates may not be available, especially outside of the United States.

For a given spatial scale, activity- and atmospheric-based approaches have sometimes been shown to lead to different conclusions, which is not unexpected as they were conceived and designed through different disciplines and for different purposes (Wilson and Swisher, 1993). Smaller-scale approaches have been shown to be effective in capturing specific emission sources and possible mitigation strategies (Duren, 2022); however, they require more resources, and emissions accounting procedures can differ temporally and from site to site (Nicholls et al., 2015). In comparison, for atmospheric-based approaches to adequately capture local-scale fluxes, a dense observational mesh is often required, though other approaches that provide less spatial information can also be used (e.g., Fitzmaurice et al., 2022; Kim et al., 2018, 2022b; Turner et al., 2020). Background observations are also needed to be able to determine finer GHG enhancements (Meijer, 2022).

Scientific research has suggested possible sources of discrepancies between activity- and atmospheric-based approaches. For example, the variation of emission factors and assumptions used in activity-based approaches has been shown to lead to overestimates of methane emissions in China inferred from inverse modeling studies (Cheewaphongphan et al., 2019). It is also worth noting that overestimates were observed to be strongly sector dependent. Other examples at the national and urban scales have shown convergence for fossil fuel CO_2 emissions, such as the work of Basu et al. (2020) in the United States where the atmospheric-based and activity-based approaches agreed to within 1.4 percent at the national scale, and work by Lauvaux et al. (2020) in the city of Indianapolis where the atmospheric- and activity-based approaches agreed to within 3 percent on a multiyear mean basis. At the facility scale, it has been demonstrated that data completeness and a mechanistic understanding of emissions estimates from each approach are important for emissions verification, and that direct comparison of emissions estimates from methods covering widely different time scales can be misleading (Vaughn et al., 2018).

Limitations of Hybrid Approaches

As discussed in Chapter 2, integration of activity- and atmospheric-based approaches and other information has the potential to offer more robust and comprehensive GHG emissions information.

The hybrid approaches inherit the strengths of activity- and atmospheric-based approaches as well as their weaknesses and limitations discussed above. Additionally, techniques used to integrate the information can also introduce challenges. For example, while data assimilation methods are generally mature, research advances (e.g., improving model resolution and mechanistic representation of processes) and work at large modeling centers have not been tailored to inform decision makers across spatial scales. In the section that follows, some additional limitations for machine learning (ML)-based approaches and the application of other digital technologies for data storage and integration are discussed in greater detail given that many of these approaches are relatively nascent and present significant potential, especially if these limitations are addressed.

Algorithmic bias. Data quality and "missingness" are primary concerns in ML applications since ML algorithms require high-quality input data to accurately train algorithms and classifiers for prediction. The lack of representative ground-truth data to validate ML-based methods is another challenge that could lead to algorithmic bias, a known problem with ML, particularly in computer vision models that have notoriously failed to accurately detect skin tone and gender (Buolamwini and Gebru, 2018; Burns et al., 2018). Validation is another main challenge for ML prediction since ground-based GHG emissions data are limited and scarcer in some regions than others. This challenge is recognized in studies that have implemented hybrid approaches for air pollution monitoring, where ground-based monitors are limited in some developing country regions, such as sub-Saharan Africa and Latin America (Shaddick et al., 2018).

Interoperability challenges. A major challenge to achieve hybrid approaches that seek to leverage and integrate the rapid proliferation of Earth observation (EO) data, nontraditional data (e.g., open social activity data), and digitally enabled collection of other relevant data that can be fused with ML algorithms is the design and development of storage, processing, and exchange systems that allow for decision makers to maximize their utility. In this domain, interoperability, or the open exchange of data between different data systems, types, and standards, is a key consideration with respect to integrating the influx of new datasets and sources while maintaining traceability and trackability (Sudmanns et al., 2020). Data taxonomies, or schemas, are critical for finding parsimony between distinct datasets and utilizing ML-based approaches for insights. Data taxonomies in other research areas such as in life sciences (with the Integrated Taxonomic Information System) and in guidelines for peer-reviewed articles (NISO, 2013) have benefited data sharing. For example, the U.S. National Research Council in 1993 first introduced the concept of a Spatial Data Infrastructure that facilitates and coordinates geospatial data exchange between systems, networks, and standards (Giuliani et al., 2017). Other examples include the Group on Earth Observation, the EU Spatial Data Infrastructure, and the Global Framework for Climate Services (Giuliani et al., 2017). Clear taxonomies allow entity matching and insight generation but will not alone combat the issues of data fragmentation (Lee and Stvilia, 2014). Federated data commons, which are a composite of diverse data sources, are a useful solution to querying heterogeneous datasets. For example, OS-Climate, a workstream of the Linux Foundation, is a data commons that unifies fragmented EO datasets using connectors like Trino. The OS-Climate data mesh is one example that allows interoperability between distinct datasets with entity matching, taxonomy, and connectors federating the dataset that would otherwise require centralized, reconfigured relational databases.

Increased energy demand. A common critique of next-generation digital technologies and data-intensive approaches to collect, process, and integrate climate data is the increased energy demand due to cloud storage and server requirements. For instance, the Centre for Environmental Data Analysis, which is the United Kingdom's atmospheric and Earth observation data center, holds over 15 petabytes of data in over 250 million files, with over 10 terabytes of data being generated and requiring storage daily (Eggleton and Winfield, 2020). Data centers are currently estimated to account for 1 percent of worldwide electricity consumption, and while there is debate in terms of exactly how much future growth in energy demand will result from new technologies, Masanet et al. (2020) found that between 2010 and 2018 energy consumption as a result of data centers only grew 6 percent due to increases in energy efficiency. Next-generation data storage, such as blockchains, as a potential decentralized climate data integration and storage solution (see Chapter 2), has also been criticized as an energy-intensive process due to the complexity of some of the cryptographic algorithms to complete computational tasks for applications like cryptocurrency (i.e., bitcoin) (Mora et al., 2018; Sedlmeir et al., 2020; Truby, 2018). The modification of distributed ledger technologies that power blockchains for data storage is an active area of research as a way to lower the transactional costs of existing modes of climate data collection while ensuring energy demands do not outpace the advantages (Schletz et al., 2022). However, there are some experiments (e.g., Microsoft[3]) to reduce energy overheads.

Data quality and uncertainty. With the velocity, variety, and volume (or three of the five Vs of Big Data identified by van Genderen et al. (2020)) of data being generated through EO and other digital technologies, a critical issue is the quality and uncertainty of the data being generated. The sheer volume of the anticipated growth in environment- and climate-related data is on the order of upward of 300 petabytes, requiring storage within the next few years (Eggleton and Winfield, 2020). Analyzing and integrating these data pose a major challenge, and even with data integration solutions such as data cubes, which aim to provide an interface for users to interact with large spatiotemporal EO data, quality is a major concern because most lack data quality or uncertainty information within metadata records (Giuliani et al., 2019). Once integrated within data cubes or other similar digital architectures, uncertainty and errors can unknowingly propagate.

Governance and transparency challenges. With the promise of digital technologies, standards to govern sharing and management of new data being generated are important considerations. Although data sharing standards such as the FAIR principles, which stand for findability, accessibility, interoperability, and reusability of data (Wilkinson et al., 2016), have been adopted as best practice for many data providers and researchers developing climate-related datasets, an evaluation of current energy databases found that many fail basic FAIR data principles that allow for interoperability and transparency (Schwanitz et al., 2022). In employing more complex, nonlinear statistical and ML approaches to improve GHG data integration, a major risk is the "black-box" problem, where such algorithms tend to diffuse information that make interpretation difficult and threaten methodological transparency (Castelvecchi, 2016).

However, the volume and complexity of data to be integrated through hybrid approaches may necessitate new governance modes and principles to ensure transparency. For example, artificial intelligence and ML approaches could introduce machine agents in lieu of human intervention requiring clearly coded rules to define authority and boundaries (i.e., smart contracts to verify data provenance) (Ramachandran and Kantarcioglu, 2017). Similarly, the emergence of decentralized autonomous organizations (DAOs), which apply blockchain technology to democratize decision making, introduces new digital collaboration and governance models with distributed ledger technology (see Box 2-7), formalizing governance rules in computer code that may potentially lower transactional costs by eliminating the need for third parties or centralized regulators (e.g., klimaDAO) (De Filippi et al., 2020; Kloppenburg et al., 2022). With these automated governance approaches, however, also come risks. Governance through DAOs is often slower moving because decisions rely on a majority vote and information asymmetries abound between members (Morrison et al., 2020).

---

[3] https://news.microsoft.com/innovation-stories/project-natick-underwater-datacenter/

4

Framework for Evaluating Greenhouse Gas Emissions Information

In the previous chapters, we introduced the need for greenhouse gas (GHG) emissions information, described the types of information that are currently available, and evaluated the strengths and weaknesses of that information. In this chapter, we present a framework for evaluating GHG emissions information for decision making. The framework is comprised of a set of pillars, or criteria, that would ideally be met in order to provide the most useful and trustworthy information to decision makers. This chapter describes the six pillars, qualitatively discusses the current capability of each approach for quantifying GHG emissions (described in Chapter 2) to satisfy the pillars, and finally presents several case studies to provide examples of how the framework could be applied to existing GHG emissions information efforts.

In developing this framework and these pillars, the Committee seeks to initiate a discussion on the approach and criteria. One of the challenges in developing a framework for assessment is recognizing the many different scales, drivers, and intended uses for GHG emissions information. Another challenge is that, unlike weather forecasting, there is limited timely and easily available feedback on the quality and accuracy of GHG emissions information. The Committee has sought to develop pillars that are mutually exclusive and collectively exhaustive. The Committee does not propose a strict hierarchy but recognizes that successful GHG emissions information will be used, trusted, and of sufficient quality to be fit for purpose.

Pillars in Establishing a Greenhouse Gas Emissions Framework

The Committee has identified six criteria or "pillars" that form a framework to evaluate GHG emissions information. The pillars can be thought of as an ideal set of characteristics that GHG emissions information should have. These pillars are usability and timeliness, information transparency, evaluation and validation, completeness, inclusivity, and communication. Each pillar is described below.

The most useful and trustworthy information for decision makers will fulfill multiple or, potentially, all pillars. Some GHG emissions information may address some pillars only partially and some pillars not at all, which is often the result of a particular activity's goals and constraints. Such information can still be highly valuable but may need to be combined with other information to enhance its usability. The objective of the pillars is to provide decision makers with a framework to qualitatively assess GHG emissions information to meet their needs. These pillars are also intended to serve as a guide for designing more useful and trusted information.

Together, these pillars provide a framework for assessing GHG emissions information in the context of its intended purpose. This means that its intended purpose should be clearly described (information transparency) and communicated. Similarly, is the information usable and timely for its intended purpose? Does the quality, reliability, and accuracy of the information meet the needs of its intended purpose? The case studies considered at the end of this chapter illustrate this approach.

Usability and Timeliness

Usability means the GHG emissions information is relevant to decision makers and that the information can be incorporated into decision making with ease. Timeliness means the GHG emissions information is available on time scales relevant to decision making. This pillar ensures that investments and efforts in GHG emissions information system development are focused on the most relevant questions, usable by decision makers, and built in such a way that meets information needs as the policy environment changes. Usability also refers to the incorporation of decision-maker input—which may be extremely useful but also has to be protected from the possibility of being inherently political—into GHG emissions information systems such that the information developed responds to the true policy needs of decision makers and stakeholders. Timeliness also means information would be available with minimal lag time and regularly updated to be relevant to policy needs.

Usability and timeliness also mean that the approach is adoptable by other users and that GHG emissions information is comparable. Comparability of the GHG emissions information with other information includes (a) comparability within the GHG emissions information (e.g., among gases, sectors, time periods) and (b) comparability among the GHG emissions information derived from different estimation methods. Approaches to achieve comparability include standardization of units, clearly differentiating fluxes versus mixing ratios, clarifying scope(s) or system boundaries, and identifying the flux unit (e.g., per unit area in a grid or "asset"). Standardization of definitions would improve comparability to other studies, while differences in grid scales and temporal coverage may be barriers to comparability. Contextualization metrics could help to increase the usability and understandability of the information.

Information Transparency

Information transparency means that the data and methods used to produce GHG emissions information are publicly available and traceable. This includes the primary and key supporting data sources and the methods or models used, including definitions, version numbers, documentation, funding sources, and open access (i.e., freely available, not behind a paywall). This pillar is related to the concept of indirect reproducibility, which assesses "the transparency of available information to allow reproducibility," and the underpinning standards of data transparency, analytic methods transparency, research materials transparency, and design and analysis transparency (NASEM, 2019). The Transparency and Openness Promotion (TOP) Guidelines developed for scientific publishing provide additional guidance on standards of transparency that could be adapted to GHG emissions information systems (Nosek et al., 2015).

The process and methods associated with source data collection and/or data transformation are also critical for information transparency. To be transparent, models, estimation procedures (e.g., conversion factors, statistical procedures, and algorithms), and analytical methods (including calibration) should be well documented and freely available wherever possible. Documentation should include details about the system boundary or domain of the calculations, types of emissions included, and key assumptions of the approach. Information transparency is essential to allow users to assess data quality and accuracy (see also the evaluation and validation pillar).

Information transparency enables trust between information providers and users, contributing to political credibility and trust between agencies and actors. Information transparency is also essential for understanding GHG emissions information and its limitations, allowing results to be independently reproduced. Finally, information transparency contributes to reliability, credibility, consistency, comparability, and calibration by way of clarity in understanding how the information is generated. A clear understanding of data sources and methods also creates opportunities for verification of data and methods, intercomparisons of alternate approaches, and innovation and application of improved methods.

Evaluation and Validation

Evaluation and validation mean that the GHG emissions information and the data and methods used to create the information have been rigorously assessed to ensure they are reliable and of sufficient accuracy and quality for the intended purpose, with uncertainties quantified. We are using the terms "evaluation and validation" to broadly capture the concept of data quality because GHG emissions information systems often combine many different types and sources of data, each with their own quality standards, and use models and transformations to convert measured data to different temporal and spatial scales with implications for accuracy. This is especially important when working with GHG emissions from one or more sources that are not normally distributed in space and/or time.

Evaluation and validation should be performed by the creators of the GHG emissions information as well as by independent auditors, researchers, or users. The Committee uses the term evaluation to refer to the quantitative or qualitative examination of the quality, accuracy, and completeness of the GHG emissions information; review of the methodologies (including calibration); traceability of data used; consideration of the appropriateness of the approach; and documentation of potential errors, biases, and uncertainties. Evaluation should be quantitative where possible. If a quantitative evaluation is not possible, a qualitative evaluation can be conducted, rather than none at all.

Validation goes beyond evaluation, is highly quantitative, and typically involves comparison to independent datasets. Validation ensures the robustness of the information and the methods and data used to produce it. Steps include testing the reproducibility of the GHG emissions information and ensuring its accuracy by comparing it to independent information where possible. Ideally, the biases and uncertainties associated with the information are quantified and a more complete assessment of the limitations of the approach is provided. Successful validation of the information ensures reliability and readiness for applications.

Completeness

Completeness is a key principle in GHG emissions information development, supporting the comparison and verification of inventories with relevant guidelines or standards. Completeness means that the GHG emissions information for the geographic boundary covers all sources and sinks and all GHGs, similar to guidelines from the Intergovernmental Panel on Climate Change (IPCC, 2006, 2019a) and other organizations. Approaches as well as information and inputs to approaches should have comprehensive spatial and temporal coverage. All emissions sources, regardless of size (i.e., no pre-defined limit on the magnitude of sources), and one-time events along with routine emissions should be included. Consideration should be made for potential errors generated when extrapolating from limited measurement campaigns to larger spatial or temporal scales to "complete" an inventory or using outdated information. There should be clear documentation of any sources that were absent, excluded, or not applicable. To achieve completeness, many different sources of information may be integrated.

Inclusivity

Inclusivity means that the GHG emissions information is generated by all relevant or applicable stakeholders, including decision makers, individual entities, locally-based experts, and the public. It considers who is involved in the creation of and covered by the GHG emissions information. Greater representation and participation would increase public understanding and trust in GHG emissions information. For example, current geographical gaps in information could be covered by satellites, airborne measurements, or even citizen science without the involvement of governments or scientists from those geographies. While this might meet scientific objectives of data coverage, it does not build support and capacity for sustained monitoring and could overlook local drivers of emissions. Similarly, while city-generated inventories may have differences from research-based inventories, the gap could be closed by capacity building and improved communication between local stakeholders and researchers.

In this context, inclusivity includes geographies and communities that have been historically underrepresented (e.g., the Global South and marginalized groups). Inclusivity also relates to participation through formal (e.g., official public comment mechanisms) and informal mechanisms. Broader public engagement can quickly help generate urgently needed data and information about emissions. Examples of public engagement include citizen science or data collection or monitoring by citizens. Deployment of innovative, low-cost sensor networks provides opportunities for broad participation in data collection, often providing some measurement data where there had been none (Kim et al., 2022b), and public review and comment allow citizens to engage in the development of plans and regulations. The more inclusive the approaches, the greater the trust can be in the process and outcomes.

Communication

Communication means that the GHG emissions information is *effectively* communicated to the public and decision makers. This communication goes beyond the information transparency pillar to consider the intended audience and "meet them where they are" in terms of communicating in accessible and understandable forms and formats the methodologies, assumptions, and their implications. To illustrate how this communication pillar complements the information transparency pillar, consider how scientific results can be published in journals or databases and fully meet the eight TOP Guidelines and yet not convey meaningful information to the public and decision makers. As another example, most people would not understand the meteorological information produced by weather services but do benefit from accessible and effectively communicated weather forecasts in the news and applications designed for the public.

Ways to effectively communicate GHG emissions information include websites that are simple to navigate, data that can be freely downloaded, data formats that are widely used (e.g., csv, json, xml), metadata that include units and other details, and documentation presented in several languages. Descriptions and supporting documentation should be provided in nontechnical language that is easy to understand. As discussed in the transparency pillar, special attention should be devoted to documenting open traceability of the underlying input/source data. Tutorials (including videos), the extensive use of visualizations, and short summaries (e.g., geographic or sectoral aggregation) of the datasets used and developed can provide useful information to nonexpert decision makers. At the same time, communicating urgency can show why the inventory is important in policy considerations. Inclusivity requires data and results to be presented and communicated in a manner that a stakeholder or decision maker can reasonably assess and understand.

Framework for Evaluating Greenhouse Gas Emissions Information

The six pillars described above provide a set of criteria that can be used to qualitatively assess GHG emissions information. The framework is flexible and can be used across a range of applications and spatial scales. Assessment of GHG emissions information using the framework will depend on the user and the need that the user is trying to address; therefore, the outcome will vary, even for the same GHG emissions information. For example, a decision maker looking to create the most effective new policies to reduce GHG emissions might need information that scores high on completeness, transparency, and communication. A decision maker looking for the most reliable assessment of the effectiveness of policy actions on a particular GHG and sector might rather emphasize usability and timeliness, evaluation and validation, and inclusivity. Various decision makers and independent bodies can apply the framework and prioritize the pillars to assess which information best suits their needs to inform the decision-making process.

To provide a high-level overview of current GHG emissions information, the Committee attempts to summarize here how the current capabilities of the general approaches perform relative to the pillars, drawing on the background introduced in Chapters 2 and 3. For each approach (activity-based, atmospheric-based, and hybrid), we evaluate below the general performance of typical current applications for both their methods and data with respect to each pillar with qualitative descriptions and scores. Methods refer to the calculations and mathematical methodologies. Data capture both the input data (e.g., activity data, emissions factors, etc.) and the resulting emissions estimates.

In applying the framework generally to each approach, we considered a broad range of users, including national decision makers, regional/local decision makers, corporate entities, various GHG emissions information creators, researchers, and the public. The needs of the users or the applications for the GHG emissions information similarly span a large range including policy design, policy compliance, and research. In our description of the performance of each approach against the pillars, we include some consideration of different users and user needs.

Figure 4-1 presents the Committee's indicative mapping of the general approaches for developing GHG emissions information described in Chapter 2 to qualitatively score for each pillar. While it is difficult to assign a score when complex and varied analyses and datasets exist for each approach, the Committee sees value in applying the framework and mapping the key strengths and weaknesses to the pillar framework. Three different qualitative scores are used to assess the general ability of the approaches to satisfy the pillars. A high score indicates that the existing approach can consistently address the pillar criteria. A low score indicates the criteria related to that pillar are not usually well addressed. A medium score indicates that the approach ranks high in some instances but ranks low in others.

These evaluations are derived from the Committees' expert knowledge and judgment, rather than through a detailed and exhaustive evaluation of a large number of programs and projects. We recognize the limitation of this strategy, but we believe there is value in making a qualitative assessment of how different methods and approaches perform relative to the pillars in order to inform key areas for improvement and additional effort. Readers should also note that we use the activity-based, atmospheric-based, and hybrid approaches in our grouping of GHG emissions information, but other aspects such as spatial scale, actors, and potential user communities are important to consider and these are also addressed in the evaluation. For activity-based approaches, we focus the scoring on national to global scale applications as outlined in the Statement of Task, recognizing

Approaches		Pillars					
		Usability and Timeliness	Information Transparency	Evaluation and Validation	Completeness	Inclusivity	Communication
Activity-based	Methods	Medium	High	Medium	Medium	Low	Medium
	Data	Medium	Medium	Low	Medium	Low	Medium
Atmospheric-based	Methods	Low	High	Medium	Medium	Low	Low
	Data	Medium	Medium	High	Medium	Low	Medium
Hybrid	Methods & Data	Low	Low	Medium	Medium	Low	Low

FIGURE 4-1 Committee's qualitative assessment of the current capabilities of three approaches for quantifying GHG emissions information against the six evaluation pillars. Methods refer to the calculations and mathematical methodologies, and data capture both the inputs (activity data, emissions factors, mixing ratio observations, etc.) and GHG emissions information outputs. Capabilities have been ranked from low (light green) to high (dark green).

that many subnational activities may have different general characteristics, which we note in the description in some cases. For atmospheric-based and hybrid approaches we consider the state of the art more broadly in the scoring.

While we identify that each general approach has typical strengths and weaknesses when viewed against this framework, the performance of any specific GHG emissions information may differ from these broad categorizations when considering the specific design, implementation, and objectives of the effort and when considering a specific user. To illustrate this specificity and how the framework can be used to evaluate an individual application, example case studies in the following section illustrate how the pillars can be used as tools to evaluate specific efforts.

Activity-Based Approaches

Usability and timeliness [*Methods: Medium, Data: Medium*]: Activity-based methods and data are ranked medium in usability and timeliness. Like most of the scoring for the activity-based approach, the volume and diversity of existing activity-based GHG information make single assignments challenging as there are exceptions in most cases.

Numerous activity-based inventory efforts have been under way for decades with new additions in recent years. Though they use different methods, they share similar barriers to methodological usability. National inventory reporting as part of the United Nations Framework Convention on Climate Change (UNFCCC) process was developed to directly assist the policy-making process and relies on guidelines and methods outlined by the IPCC (IPCC, 2006). Numerous inventories developed within the academic and research community rely on the general methodological guidance of the IPCC but often develop additional, specific methodologies for particular questions of interest. All inventories publish accessible methodologies (of varying thoroughness) but in forms and mediums often requiring expert knowledge. Furthermore, because many inventories were not initially targeted at the decision-making community, methods and methodological descriptions are not always useful to decision makers where simple emission metrics, cross-comparability, and space/time resolutions are focused on governance and/or land-use boundaries. There is a lack of standardization of core methodological elements such as units, system boundaries, and geographic boundaries that would enable comparability across products. The development of inventories with higher information granularity is often time and labor intensive and, unlike national inventories, they may not be updated to regularly reflect updated activity data. Therefore, the use of the activity-based approach has typically been limited to a small, technical community because of its complexity and highly technical language.

The timeliness of the methods and data has undergone significant change in the last few years with the pressure placed on inventories to respond to the COVID-19 pandemic with more updated information (Le Quéré et al., 2020). Indeed, some global GHG inventories are producing GHG information with latencies of a few months (e.g., Dou et al., 2022; Forster et al., 2020; Jackson et al., 2022; Liu et al., 2020b), though they may not be as robust or as complete, and lack thorough evaluation (Oda et al., 2021). This pertains to data inputs or outputs and methodological documentation. However, the majority of activity-based GHG inventories have latencies of 2 to 3 years, often driven by the latency of data sources (e.g., updated activity data) which take time to produce and perform quality assurance and quality control.

We rank the usability of the activity-based data as medium. Activity-based GHG emissions information is often the primary data source used by decision makers, including the UNFCCC, national and subnational policy makers, corporations, and the public. However, the usability of activity-based GHG emissions information can be limited—depending on the application—by latency, GHG species coverage, a lack of spatial and temporal information that may be required by stakeholders, and comparable units and definitions. For example, 121 countries have signed the Global Methane Pledge which aims to reduce global methane emissions by at least 30 percent from 2020 levels by 2030 with a potential of reducing 0.2°C warming by 2050.[1] Yet, methane emissions data remain uncertain or missing from some UNFCCC national submissions, which in turn impacts policy and scientific goals, and the ability to verify progress. In this light, it is notable that signatories to the Global Methane Pledge also "commit to moving towards using the highest tier IPCC good practice inventory methodologies, as well as working to continuously improve the accuracy, transparency, consistency, comparability, and completeness of national greenhouse gas inventory reporting under the UNFCCC and Paris Agreement, and to provide greater transparency in key sectors."

Information transparency [*Methods: High, Data: Medium*]: Global/national GHG inventories typically have well-defined, transparent, and documented methodologies, such as those following the IPCC guidelines (IPCC, 2006) or the detailed documentation found in nearly every global inventory produced within the research community. It is worth noting that at smaller scales (e.g., urban, facility, corporate), methodologies are often not fully reported. Nevertheless, for the global GHG emissions information efforts, methodological transparency is scored as high.

Regarding input data, some global/national data sources are publicly available but many are not, even if they are clearly identified (e.g., energy data). This is particularly important for the energy-related emissions which rely on a common, small set of national energy datasets. The emissions estimates that are produced by global inventory efforts are generally available and easy to access for global/national datasets (e.g., national reports for the UNFCCC; emission estimates based on primary energy data sources from the International Energy Agency or BP; compiled emissions datasets, for example, Open-source Data Inventory for Anthropogenic CO_2 or the Global Carbon Project), but they can be in a wide variety of formats (see communication discussion). Overall, transparency in data and methods tends to be higher for global and national information but lower for facility- or city-scale information, and global aggregates of these. We scored the information

---

[1] https://www.globalmethanepledge.org/. As of Sept 1, 2022.

transparency for activity-based data as medium, given the current mix of transparency for input versus data outputs.

Evaluation and validation [*Methods: Medium, Data: Low*]: Of the activity-based inventories described in Chapter 2, the UNFCCC GHG information methodology undergoes a unique codified administrative evaluation (e.g., auditing the adherence to IPCC guidelines). Similar efforts are in place for the global aggregation of facility-scale information. However, for all global inventories, methodological evaluation such as comparisons of multiple methods, cross-checks of overlapping data, and documentation of potential errors and biases exists in some cases but is generally limited. Therefore, we scored the evaluation and validation for activity-based methods as medium. Evaluation and validation of input data and data products is similarly limited though there has been increasing work on activity-based intercomparisons (Han et al., 2021; Hutchins et al., 2017; Oda et al., 2018) and cross-approach intercomparisons (Deng et al., 2022). Such efforts provide important near-term steps toward evaluation, particularly if results are traced to methods or input data and recommendations for future improvements are made. Nevertheless, comparison to independent data and assignment of numerical uncertainty are not consistently performed (with some recent exceptions) and these are key goals in evaluation and validation of GHG emissions estimates. As a result, we scored the evaluation and validation for activity-based data as low.

Completeness [*Methods: Medium, Data: Medium*]: Completeness of activity-based methods refers to methodologies that are designed to trace all GHGs and all emission sources and sinks, capture one-time events along with routine emissions, and include emissions of all magnitudes. Activity-based methods often include major gases and sources as part of the methodological intent, although this varies. For example, many global inventories are focused on carbon dioxide (CO_2) only while others methodologically tackle all gases, even if data support is limited in the production of estimates. Furthermore, there is variation in methodological completeness across sectors or data sources. Inventories can mix algorithmic thoroughness by using new or novel methods to estimate particular emitting source categories but inherit results from separate inventories for others. There can also be gaps in methods, such as from the use of a minimum threshold for the magnitude of sources included in an inventory and the inability to capture nonroutine emissions. Many of the research community's activity-based estimates are not regularly updated and thus do not provide temporally completed emission estimates.

As with the methodological limits, the input and output content of global inventories varies widely from single gases or sectors, often estimated with considerable detail and thoroughness, to more comprehensive efforts. Input data sources are often "sole source," not availing of the potential for overlapping data sources or data constraints that exist at more than one space or time scale. Exceptions exist where multiple sources of overlapping data are employed in emissions estimation. Completeness may not be achieved when limited measurement campaigns are extrapolated for broader applications. Overall, we scored the completeness for activity-based methods and data as medium.

Inclusivity [*Methods: Low, Data: Low*]: Methodological development (including calibration procedures) and input and output information are typically built and gathered by experts. It is worth noting that at local scales, where interaction and communication are potentially easier, there may be more collaboration on methods and data with users and stakeholders. However, at the global/national scale, public interaction and comment opportunities could engage a wider array of participants than is current practice. The UNFCCC process is an exception to this as it does involve large amounts of interaction with the international negotiating community directly and indirectly insofar as they represent a wider collection of interests. However, there remains considerable untapped potential to increase inclusivity via stakeholder engagement, citizen science efforts, and iterative surveys and outreach efforts. The limitations in inclusivity relate directly to the usability and timeliness pillar in that more and earlier interaction with the user community can better inform aspects of usability currently missed in the existing information domain.

Communication [*Methods: Medium, Data: Medium*]: Communication of activity-based methods and data spans a wide range. Global/national GHG inventory methods are typically communicated to the target decision-maker communities, though communication to decision makers at smaller scales is not emphasized. For example, UNFCCC inventory information is often further collected and summarized by the nongovernmental organization community to enhance understanding and use by both international decision makers and other communities such as the media, citizen organizations, and the business community. Some of the research-based inventories develop websites and other communication platforms, but these are more often targeted to expert audiences. Exceptions include some programs that have created effective, easy-to-understand descriptions and infographics for decision makers and the public. In general, methodological communication remains limited and not well suited to the decision-making community. Methodological communication along with the inclusivity and evaluation and validation pillars are critical elements in building the trust of information outputs. Hence, both the methodological and data communication is scored as medium, reflecting the wide variation in both communication focus and quality.

Atmospheric-Based Approaches

Usability and timeliness [*Methods: Low, Data: Medium*]: Usability and timeliness in atmospheric-based approaches can vary. Often, due to the complexity of the methods and results, coarseness of the emissions estimates, and lack of specific attribution of emissions sectors, the usability of atmospheric-based methods and data can be low. On the other hand, the data from some approaches are highly usable, such as estimates of point-source emissions. Timeliness varies, as some global-scale methods require emissions estimates as well as meteorological data and measurements that have a time lag of years. In more recent work, researchers have sought to provide timelier analysis, though time lags of at least a year or two are still typical and may be longer for some regions. Some collaborative efforts have sought to improve the usability of the data from atmospheric-based approaches by improved communication between policy makers and researchers (e.g., European Union's [EU's] VERIFY project (Peylin, 2022), National Aeronautics and Space Administration's Carbon Monitoring System [Hurtt et al., 2022]). Some collaborative efforts between researchers and policy makers exist, such as work in the states of Colorado and California and U.S. Environmental Protection Agency (EPA) collaborations with the research community (Duren et al., 2019; Fischer et al., 2020; Graven et al. 2018). Often the conclusions of the studies, rather than the specific numerical estimates of emissions—particularly after the synthesis of consistent findings by multiple studies—can be highly usable by decision makers. For example, multiple studies showing the presence of methane "super-emitters" and consistently higher natural gas leak rates from certain infrastructures compared to standard emission factor approaches (Alvarez et al., 2018) have helped to direct government and corporate priorities for methane mitigation actions in the oil and gas sector. Overall, we scored the methods as low and data as medium for usability and timeliness.

Information transparency [*Methods: High, Data: Medium*]: Atmospheric-based methods are generally well documented in the academic literature and public websites. For this reason, we rank the transparency of atmospheric-based methods as high, although we note that costs to access some academic sources may be a barrier in some cases. The complexity of some methods may also present a barrier to usability as described above. The input data and derived emissions estimates or flux estimates are described in the peer-reviewed literature in nearly all cases, but the numerical input data and estimates are not always provided, so we ranked information transparency for data as medium. Recently, many journals have instituted requirements for accessibility of input data, results, and model code, further improving accessibility and reproducibility.

Evaluation and validation [*Methods: Medium, Data: High*]: A strong point of atmospheric-based approaches is that evaluation and validation of some aspects of measurement methodologies are well integrated with the technique and are supported by peer-reviewed descriptions. However, other aspects of the methodology are difficult to evaluate, including model transport uncertainty. For data and results of atmospheric-based approaches, there is generally validation and a quantitative assessment of uncertainty. Observations are typically calibrated and reported following standard practices (see Box 2-5). There are also activities regularly comparing and synthesizing different atmospheric-based results (e.g., TransCom-CH_4, Global Methane Budget), which further support the evaluation and validation of the data. As a result, we rank atmospheric-based methods as medium (primarily due to the models used to convert measurements to emissions) and data as high for evaluation and validation.

Completeness [*Methods: Medium, Data: Medium*]: Atmospheric-based approaches are typically comprehensive, as all sources and sinks are captured by the nature of atmospheric measurements that integrate fluxes. However, attributing GHG sources or sinks to specific sectors is a difficult task, so information is often complete in terms of total fluxes but incomplete in terms of specifying source sectors. Alternately, assumptions about the distributions or magnitudes of some sources can lead to biases in the resulting GHG source or sink estimates. The use of tracers to attribute sources is important for enhancing this type of completeness. Atmospheric analyses are frequently focused on single gases; multigas analysis is less common. Spatial and temporal coverage varies, as some information is based on limited measurement campaigns in specific regions while others are global. The spatial and temporal resolution can be coarse, with average emissions over regions and months or years commonly reported. Some techniques have a high temporal resolution in a fixed location (e.g., plume detection, eddy covariance). Therefore, we rank the atmospheric-based methods and data completeness as medium, given that they sense all sources, but frequently with only single-gas analysis, limits on spatial and temporal information, and difficulty separating sources and sinks.

Inclusivity [*Methods: Low, Data: Low*]: The inclusivity of atmospheric-based methods and data is typically low, with little input from stakeholders on the methodology and data collection. Method development, atmospheric measurements, and data analysis are primarily performed by a small, technical community. In some regional and local work, there may be more collaboration with some measurements collected by corporations, local agencies, or citizens with engagement between technical experts and decision makers (see communication discussion below). Nevertheless, we ranked both methods and data low for the inclusivity pillar as these practices are more exceptional than standard.

Communication [*Methods: Low, Data: Medium*]: Method communication is typically low for atmospheric-based approaches, as the primary venue for the presentation and discussion of this work is in scientific literature and specialized conferences, and thus not communicated widely. While methods are documented, they are typically in technical language and are quite complex. Communication of results can vary. The data and results from these approaches are often reported in scientific publications and technical conferences, although some efforts have been effectively communicated to decision makers and have had strong impacts on policies or corporate activities (e.g., natural gas leaks, the Kigali Amendment, CFC-11 detection). Sometimes efforts to communicate with decision makers and the public through the media, such as when journalists present a new finding, can be very effective. As mentioned in the usability discussion above, some efforts have aimed to improve usability via collaboration between scientists and decision makers, which also helps in the communication of atmospheric-based efforts to users. In light of this, we ranked the communication of atmospheric-based methods as low and data as medium.

Hybrid Approaches

Hybrid approaches that integrate different approaches and methodologies, potentially using advanced computational tools, are advancing and developing rapidly. They have the potential to capitalize on the different strengths and compensate for weaknesses of individual approaches, as well as innovate new approaches. While doing so, it may be a challenge for hybrid approaches to ensure the pillars of transparency, inclusivity, communication, and usability are addressed.

The methods and data of the hybrid approach are closely entwined, so we ranked the combined methods and data against the pillars.

Usability and timeliness [*Low*]: Usability and timeliness is currently ranked low, although there is potential for excellent performance against this pillar with the hybrid approach. The low ranking is driven by the fact that this work is still in development in a small community of experts, and the demonstration of relevance and usability by decision makers is not well established. The hybrid approach has the potential to provide timely data and, if developed with attention to standardization of definitions and units and the key needs of decision makers, it could score very well against this pillar.

Information transparency [*Low*]: We scored the hybrid approach as low for information transparency. It has many of the characteristics of the atmospheric-based approach, with complex methods, the need for expert knowledge, and some limits of access to information with paywalls and unreleased data and methodologies that are not always well documented. For example, a lack of transparency generally exists in the private-sector development of hybrid approaches, which may combine machine learning and other digital data collection technologies and are not peer reviewed or open source.

Evaluation and validation [*Medium*]: Evaluation and validation is ranked medium, as the hybrid approach should include evaluation and validation as part of the development and peer review of the methods. However, some aspects of the approach may be difficult to evaluate. Data products may be compared to other results, although not all products have independent evaluation.

Completeness [*Medium*]: Completeness is ranked as medium. The integration of data, potentially including atmospheric measurement data, activity data, and ancillary information, aims to provide a robust and complete picture, with the potential to rank high. However, it may be challenging to address all gases and sectors with a high degree of data coverage and data constraints. The hybrid approach may have limitations in the separation of sectors but is likely less limited than atmospheric-based approaches.

Inclusivity [*Low*]: Inclusivity is scored as low, as this work is currently conducted by a small group of experts with methods that are not uniformly disseminated widely.

Communication [*Low*]: As with usability and timeliness, communication is ranked as low for the current generation of hybrid approaches, given the limited dissemination of results and details. So far, much of this information is in conference proceedings, technical journals, or hard-to-access reports. On the other hand, some emerging hybrid approaches could address these pillars well at the outset by creating open-source software, publicly available datasets, community engagement, and web-based tools.

Case Studies Applying the Framework

The six case studies that follow provide examples of activity-based, atmospheric-based, and hybrid approaches for quantifying GHG emissions across a range of scales. Each case study is qualitatively evaluated against the six pillars described above and areas for improvement are identified. The evaluation relative to the pillars is done with multiple users in mind, and we seek to indicate where the pillars are met for one community (e.g., researchers) but not another. These case studies demonstrate how the framework could be used to evaluate current efforts and identify their strengths and weaknesses.

Case studies were selected to illustrate the range of approaches outlined in this report, including the improvements to the well-established activity-based approach of the UNFCCC in the Enhanced Transparency Framework (ETF), an atmospheric-measurement-driven effort (Permian Methane Analysis Project), and several mixed or hybrid approaches. We sought to examine projects that used a range of approaches and that also exist in different communities—the policy world, research community, cross-community efforts—and across a range of scales. The set of case studies is only a small sample of current activities, but they provide a glimpse into the wide range of existing efforts and demonstrate how the framework can be applied to evaluate specific GHG emissions information. From these case studies, we see that innovation in atmospheric-based and hybrid approaches can improve traditional activity-based approaches and support target-based mitigation strategies.

4.1 UNFCCC Enhanced Transparency Framework

In this case study, we consider the ETF, established by the 2015 Paris Agreement, for country-level reporting to the UNFCCC. The ETF includes requirements for both developed and developing countries to report on various climate-related actions, implementation, and support (UNFCCC, 2021b). As part of the ETF, all Parties are required to report, on a biennial basis, a national GHG inventory according to the IPCC 2006 guidelines and report on progress made in implementing and achieving their nationally determined contributions (NDCs) (UNFCCC, 2015, Article 13). The GHG emissions inventories are to be reported electronically using the common reporting tables. These reporting requirements are one of the few legally binding commitments for both developed and developing Parties in the Paris Agreement (Weikmans and Gupta, 2021). Both groups of countries are required as part of their Biennial Transparency Reports to include a GHG emissions inventory, following the same requirements in the modalities, procedures, and guidelines, with some specific flexibility given to those developing countries who need it in light of their capacity. The first Biennial Transparency Reports are due at the end of 2024, with some discretion in timing for Small Island Developing States and Least Developed Countries. To meet current requirements under the Convention, developed countries must also continue to report an annual GHG inventory in years in which a Biennial Transparency Report is not due. Recognizing the various capacities of developing countries, the Global Environment Facility and the UNFCCC's Consultative Group of Experts provide support for developing inventories (UNFCCC, 2020).

In the evaluation below (see also Figure 4-2), we consider users of the national reports and inventories from the ETF to be national decision makers as well as other users such as researchers or the public.

Usability and timeliness [*Medium*]: For the UNFCCC user community, usability and timeliness is high because reporting is standardized, approved, and used at the international level. For other users, including the research community and corporations, usability and timeliness of data is lower, as the 2024 reporting will cover emissions from 2022 employing 2006 IPCC guidelines which may not include updated findings, and the information or information formats may not be optimal for their uses. Therefore, usability and timeliness is scored medium.

Information transparency [*High*]: The new provisions under the ETF place a greater emphasis on reporting as the key mechanism of the Paris Agreement's "pledge and review" approach where the rules of the Paris framework are envisioned to make countries' actions transparent and to hold countries accountable to their NDCs (Gupta and van Asselt, 2019). Countries are required to report according to the documented IPCC 2006 guidelines. If countries all report and meet the

Case Study	Pillars					
	Usability and Timeliness	Information Transparency	Evaluation and Validation	Completeness	Inclusivity	Communication
UNFCCC ETF	Medium	High	Medium	High	Medium	Medium

FIGURE 4-2 Qualitative evaluation of the United Nations Framework Convention on Climate Change (UNFCCC) Enhanced Transparency Framework (ETF) against the six pillars.

requirements, transparency should be high. However, this is currently not true for all countries, so transparency will depend on the extent to which countries adhere to the 2006 guidelines.

Evaluation and validation [*Medium*]: Under the ETF, all countries are subject to a review of their GHG inventory by an expert review team who will publish a report of their findings on the UNFCCC website. All members of the review team must take and pass a training course to be certified to conduct the review. Under the modalities, procedures, and guidelines, Parties are encouraged, but not required, to undergo a separate peer review. Other opportunities for evaluation and validation, such as comparison to independent information and estimation of uncertainties, are likely to be done by only a few countries. The use of Tier 1 methods and generic or out-of-date emission factors may reduce the accuracy of the resulting estimates. Therefore, evaluation and validation is scored medium.

Completeness [*High*]: The ETF is likely to score highly for completeness since countries will follow the IPCC 2006 guidelines. However, the guidelines allow for flexibility in the level of detail in national GHG inventories, including the scope of the inventory's sectors and gases, among other dimensions (Goodwin and Kizzier, 2018). Therefore, some countries' reporting may be low or medium for completeness.

Inclusivity [*Medium*]: Inventory development and reporting will be inclusive at an international level as it will be conducted by all nations that are Parties to the Paris Agreement. However, nation-level reporting will be done at the political level and may not include all stakeholders. The process typically remains limited to a small group of technical experts, with little community engagement. Some developing countries have limited experience with the common reporting tables and may need capacity building (Falduto and Wartmann, 2021). Furthermore, underlying activity-based information may not be globally representative; for example, the IPCC Emission Factor Database was created using data from Organisation for Economic Co-operation and Development countries. Therefore, inclusivity is scored as medium.

Communication [*Medium*]: The inventories that the Parties report to the UNFCCC and the common reporting tables will be available for download on the UNFCCC website. This will provide a uniform and familiar interface for people engaged in the UNFCCC process. For the general public and research communities, there is very limited effective, accessible, and understandable information provided by the UNFCCC directly. As a result, communication is ranked as medium.

4.2 Research Community's Estimate of China's National Emissions Based on Activity Data

This case study considers the work of Shan et al. (2020) as an academic research study to estimate China's national GHG emissions using an activity-based approach at a finer level of detail than is available from the official UNFCCC reporting. In China, GHG emissions inventory information has been officially reported in its National Communications and Biennial Update Reports for 1994 (NDRC, 2004), 2005 (NDRC, 2012), 2010 (NDRC, 2019b), 2012 (NDRC, 2016), and 2014 (NDRC, 2019a). Since these reports have been released sporadically and lack detail and access to the underlying raw data, transparency is low. Shan et al. (2020) sought to assemble a detailed and transparent CO_2 emissions inventory for China for fossil fuel and cement production, in which more country- and region-specific information was used than global inventory estimates for China such as EDGAR (Emissions Database for Global Atmospheric Research) and CDIAC (Carbon Dioxide Information Analysis Center).

Parallel efforts in the government and the research community to estimate GHG emissions occurs not just in China, but is widespread in many developed as well as developing countries (see Case Study 4.4 below for another example). The work in China by Shan et al. (2020) is a particularly interesting case study because China is presently the largest contributor to global GHG emissions and prior work has revealed discrepancies of ~18 percent in GHG emissions calculated with different sources of activity data or emission factors, including national versus provincial level activity data and coal characteristics (Guan et al., 2012).

Shan et al. (2020) used national and provincial inventories and emissions factors from Shan et al. (2018) and Liu et al. (2015). In comparison to Guan et al. (2012), Shan et al. (2020) found that the activity data have improved over time, although both the activity data and emissions factors are important contributors to uncertainty. They applied the IPCC's recommended Monte Carlo technique to estimate uncertainties on the overall emissions inventory and they report an uncertainty range for energy-related emissions of –15 percent to +30 percent at 97.5 percent confidence for 2016–2017. To create a transparent and verifiable dataset, the authors published the activity data, emission factors, and calculation code with the inventories (Shan et al., 2020). The Shan et al. (2020) study highlights how efforts in the research community can evolve rapidly, in parallel with ongoing reporting within the UNFCCC framework, and offers additional insights into the contributors to and trends in uncertainty. These research efforts provide access to valuable underlying datasets and produce inventories that can be important inputs to other analyses, including atmospheric-based studies.

In the evaluation below (see also Figure 4-3), we consider GHG emissions information available from the Shan et al. (2020) study and users of the information to include national decision makers, researchers, and the public.

Usability and timeliness [*Medium*]: The data are in a structure that is likely to be very usable to national decision makers and researchers, using the 47 sectors that are in the Chinese energy statistics system. The units are clearly identified, and the data have been compared to other estimates. However, specific data products or descriptions of results that are aimed toward policy makers or the public have not been created or disseminated. Broader usability may be limited by the use of sectors that may not align with the IPCC sectors. The Shan et al. (2020) paper reports 2016 and 2017 emissions, a time lag that is important in a place with rapidly changing emissions and industries. Timeliness could be improved to enhance usability by decision makers and the public. Considering all these factors, a score of medium is given for usability and timeliness.

Information transparency [*High*]: Data and methods from the study are documented and freely available online, after user registration. However, it is important that data access is maintained over time as some data from individual research studies such as this may become inaccessible if not deposited into a reliable data center. Therefore, transparency is rated high.

Case Study	Pillars					
	Usability and Timeliness	Information Transparency	Evaluation and Validation	Completeness	Inclusivity	Communication
Shan et al. (2020)	Medium	High	High	Low	Low	Medium

FIGURE 4-3 Qualitative evaluation of national activity-based emission inventories from Shan et al. (2020) against the six pillars.

Evaluation and validation [*High*]: The peer-reviewed papers provide a comparison of their datasets with a wide range of other inventories to validate the findings. The authors evaluated the data sources, applicability of emission factors, documented potential errors, and calculated uncertainties. Overall, this effort is ranked as high for evaluation and validation but could be improved further with comparison to independent data, including atmospheric GHG data.

Completeness [*Low*]: Only CO_2 emissions for certain sectors are estimated without estimating other GHGs or sectors, so completeness is ranked low.

Inclusivity [*Low*]: Inclusivity is ranked low since this work was conducted by a small group of researchers and did not appear to engage with individual entities, locally based experts, or the public.

Communication [*Medium*]: The data are downloadable and are in a widely used format (csv), but are likely to reach a small audience already familiar with such inventories in the research community. The descriptions of the data and methods are provided in an academic research paper, which are not likely to be easily discoverable or understood by decision makers or the public. There is no identified communication to the public and policy makers.

4.3 Urban Hybrid Approach: INFLUX

In this case study, we consider the development and use of a hybrid approach for estimating GHG emissions on an urban scale: the Indianapolis Flux Experiment (INFLUX) in the U.S. city of Indianapolis (Davis et al., 2017; Whetstone, 2018). Though not a global study, it offers a series of helpful clarifications of the framework in an experiment that has scaling potential in the future. Numerous academic institutions, researchers, and federal science agencies combined efforts to integrate state-of-the-art activity-based GHG estimation with atmospheric-based approaches. Initiated in 2010, INFLUX encompassed high-resolution (i.e., building, street-segment scale) activity-based estimation of fossil fuel CO_2 emissions (the Hestia Project), atmospheric mixing ratio measurements of carbon monoxide (CO), methane, and CO_2 at the surface (mobile and stationary), atmospheric $^{14}CO_2$ measurements at surface locations, aircraft measurements of CO_2 and methane fluxes, and upward-looking Total Carbon Column Observing Network column CO_2 measurements (Davis et al., 2017; Turnbull et al., 2019; Whetstone, 2018). To support atmospheric inverse modeling, state-of-the-art Eulerian/Lagrangian transport models supported by a suite of new meteorological measurements were deployed (Lauvaux et al., 2016).

The initial aim of INFLUX was to better understand the many methodological challenges and uncertainties associated with activity-based and atmospheric-based approaches to estimating CO_2 and methane emissions at urban scales. Additionally, INFLUX has progressed the furthest toward a hybrid framework through simple iteration and adjustments between the activity- and atmospheric-based approaches and combined different atmospheric-based techniques (tracer-tracer, eddy-flux, mass balance, inverse methods) to inform, for example, sector partitioning (Nathan et al., 2018; Turnbull et al., 2019; Wu et al., 2022). The experiment has significantly advanced knowledge on the more complete integration of approaches with numerous papers examining aspects of the modeling and measurements in addition to a direct comparison to the practitioner-based estimate made by the city (Mueller et al., 2021). While the initial aim was to explore methodological advances, it has shown surprisingly close convergence between the state-of-the-art atmospheric- and activity-based estimation with agreement within 3 percent over a multiyear timeframe (Lauvaux et al., 2020).

Though not necessarily the intention of INFLUX, it offers an example of a scientifically rich study that has provided GHG emissions information to the scientific community but has not devoted necessary resource investment toward transforming the GHG emissions information into forms usable by decision makers or incorporated stakeholders into the initial design and approaches to the research.

In the evaluation below (see also Figure 4-4), we consider GHG emissions information available from both the activity-based effort (the Hestia Project) and the atmospheric-based effort described and cited in Whetstone (2018) and Davis et al. (2017) as it related to users of the information including local decision makers, researchers, and the public.

Usability and timeliness [*Low*]: The methods, data, and findings of INFLUX are highly usable by the research community. However, specific data products or descriptions that are aimed toward policy makers or the public have not been created or disseminated in ways that encourage uptake. The results have not been used by or institutionally embedded in any of the decision-maker communities. Timeliness is low, as the experiment does not produce emissions estimates regularly but is primarily associated with analysis papers. However, it is worth noting that data associated with atmospheric observations are maintained with little latency. To improve usability, engagement with local policy makers and the public from the beginning of the project would have ensured the research addressed key questions and communicated results in useful and timely ways. Considering these factors, usability and timeliness is rated low, since the only group that can (so far) use the outcomes of the studies is the research community.

Information transparency [*High*]: As with the Shan et al. (2020) study in the previous section, the data and methods from the INFLUX study are documented and freely available online (e.g., Gurney et al., 2012; Heimburger et al., 2017; Miles et al., 2017). Therefore, transparency is rated high.

Evaluation and validation [*High*]: As the primary aim of INFLUX was the development and evaluation of urban-scale GHG estimation by various techniques, there has been extensive comparison of multiple approaches, as well as overlapping data, sensitivity studies, and uncertainty estimation (e.g., Deng et al., 2017). INFLUX exemplifies a very extensive implementation of evaluation and validation that is unlikely to be feasible in many cases. Therefore, evaluation and validation ranks high.

Completeness [*High*]: INFLUX produced estimates of anthropogenic CO_2 and methane emissions (and natural fluxes of CO_2). It considered all sectors without implementing thresholds or omitting nonroutine emissions. The analysis encompassed several years of measurement and estimation. Since CO_2 and methane are responsible for the vast majority of emissions in Indianapolis, an overall ranking of high is given here, while completeness could be improved further by consideration of additional GHGs or extending the system boundaries to Scope 2 or even Scope 3 emissions.

Inclusivity [*Low*]: INFLUX was conducted by a group of technical experts and researchers. There was limited engagement with practitioners and citizens in the city (e.g., local utility data were acquired), so inclusivity is rated low. Engagement with local policy makers and the public from the beginning of the project would have ensured that local knowledge was incorporated and potentially incorporated citizen science to broaden the measurement campaigns even further.

Communication [*Low*]: Data are publicly available but not in forms that are easily used or understood by local practitioners, policy makers, or the public. Key findings have been discussed within the research community and presented in academic research papers, but results in non-technical language have not been disseminated.

Case Study	Pillars					
	Usability and Timeliness	Information Transparency	Evaluation and Validation	Completeness	Inclusivity	Communication
INFLUX	Low	High	High	High	Low	Low

FIGURE 4-4 Qualitative evaluation of the Indianapolis Flux Experiment (INFLUX) against the six pillars.

4.4 Methane Emissions Estimates from the U.S. Oil and Gas Supply Chain

In this case study, we consider the estimation of methane emissions from U.S. oil and gas facilities using different approaches. The motivation for including this case study is the increasing evidence from multiple studies that methane emissions from the oil and gas sector, particularly from production facilities, are underestimated in official estimates reported by the United States and other countries (Alvarez et al., 2018; Lu et al., 2022). Recent peer-reviewed research has shown the impact of high-emitting episodic activities, unaccounted emissions sources, and outdated emission factors on overall emissions (Chen et al., 2022b; NASEM, 2018; Schwietzke et al., 2017; Vaughn et al., 2018).

Innovation in methane emission quantifying technologies can support the development of measurement-informed inventories. While not without uncertainty, these measurement-informed inventories can improve the quality in terms of completeness and more representative GHG emissions data compared to existing national activity-based methodologies. The increasing deployment of novel technologies to quantify methane emissions and develop measurement-informed inventories, if developed at scale, can improve the timeliness and accuracy of GHG emissions estimates including supporting the U.S. GHG national inventory (GHGI) and U.S. EPA Greenhouse Gas Reporting Program (GHGRP) inventory development, international programs such as the International Methane Emissions Observatory, and support target-based mitigation strategies that are cost-effective and externally credible. The recently passed Inflation Reduction Act (IRA) and the CHIPS and Science Acts focus on the development of empirical GHG measurements. The IRA explicitly directs the U.S. EPA to revise the GHGRP emissions data with empirically derived methane emissions data at an individual-facility level. The inter- and intra-day variability in methane emissions can potentially be addressed by pairing high-frequency snapshot measurements with continuous emissions monitoring systems and a mechanistic understanding of facility-level operations, in addition to expert evaluation (Wang et al., 2022a). Such measurement-informed inventories and transparent reporting can also address the limitations of activity-based methods in capturing unaccounted or poorly characterized emission sources.

Here, we consider the activity-based approach used by the U.S. EPA and an independent regional assessment called the Permian Methane Analysis Project (PermianMAP) led by the Environmental Defense Fund (EDF) to estimate methane emissions from U.S. oil and gas facilities in the Permian basin. For the U.S. EPA, new scientific studies can be used directly to update the inventory methods and emission factors if the studies are representative and geographically diverse.[2]

4.4.1 EPA estimates

In its national GHGI, the U.S. EPA estimates that the oil and gas sector is the largest source of anthropogenic methane emissions in the United States, accounting for about a third of the total methane emissions in 2020 (EPA, 2022). In 2009, the U.S. EPA finalized the annual reporting of GHGs, codified under 40 CFR Part 98 of the GHGRP from large facilities emitting over 25,000 metric tons of CO_2 equivalent, including oil and gas. Both the GHGI and GHGRP programs generally follow the 2006 IPCC Guidelines and the results are made publicly available on the U.S. EPA's website.[3] Detailed explanations of the U.S. EPA methods for estimating methane emissions from petroleum and natural gas systems are provided in each inventory report and data files are available

---

[2] Revisions and Confidentiality Determinations for Data Elements Under the Greenhouse Gas Reporting Rule, April 29, 2022. Docket Id. No. EPA-HQ-OAR-2019-0424.
[3] GHG inventory data are available at https://www.epa.gov/ghgemissions/inventory-us-greenhouse-gas-emissions-and-sinks-1990-2020. GHGRP information at a facility level can be found at https://ghgdata.epa.gov/ghgp/main.do.

to download.[4] The U.S. national GHGI undergoes a 30-day expert review. Additionally, both inventories follow a 30-day public review after expert review through publication in the U.S. Federal Register. The U.S. EPA incorporates the feedback as deemed appropriate and publishes responses to the comments when the inventories are finalized and submitted to the UNFCCC. The GHGRP follows regulatory-approved accounting methods, including default emission factors, which were finalized after public review and comment. The U.S. EPA inventories are the basis for U.S. NDCs and historical baseline emissions (Executive Office of the President, 2021). Public policies related to mitigation of GHG emissions employ the U.S. EPA inventories (EPA, 2021).

Multiple studies employing both activity- and atmospheric-based measurements have concluded that the U.S. EPA or other activity-based inventories in the United States and other countries underestimate methane emissions from the oil and gas sector (Lu et al., 2022). The 2018 National Academies study attributed the gap between conventional activity- and atmospheric-based approaches to include the persistence of high-emitting episodic activities, unaccounted or poorly characterized emission sources, and outdated emission factors (NASEM, 2018). Recent research has shown the impact of such episodic events and emission sources on overall emissions estimates (Chen et al., 2022b; Cusworth et al., 2021b; Omara et al., 2022; Schwietzke et al., 2017; Vaughn et al., 2018).

In the evaluation below we consider GHG emissions information generated by the GHGI and GHGRP and users of emissions information from oil and gas facilities, including national decision makers, researchers, and the public.

Usability and timeliness [*Medium*]: Usability and timeliness is medium. GHGIs are used for the U.S. NDCs and a historical baseline, so they are highly usable at the national level. Regional or gridded data are not currently available, which limits usability by local and regional policy makers and researchers. The latest GHGI submitted in April 2022 covers methane emissions for the period 1990–2020. While methane emissions estimates are filed by affected facilities under regulatory mandates of the GHGRP by March 31 for the prior calendar year, public access to the data lags by about 18–24 months. The GHGRP data are analyzed and integrated as appropriate into the GHGIs for the petroleum and natural gas sector.[5] The requirements under the IRA to update U.S. EPA emission factors with empirical or measurement-informed data will improve the usability of the GHGRP and GHGI. However, the time lag in the availability of the methane emissions data limits planning and decision making.

Information transparency [*High*]: The U.S. EPA national inventories can generally be considered to rank high in transparency because the methods are provided in each inventory report and the data files are available to download. The U.S. EPA follows regulatory-approved accounting methods, including default emission factors, which were finalized after sharing for public review and comment. Webpages include information about the inventories that is targeted at nonexpert audiences.

Evaluation and validation [*Medium*]: Evaluation and validation is ranked medium. GHGIs go through a 30-day expert review and the methodology is validated, though comparison to independent data sources has revealed differences between measurement-based inventories and the U.S. EPA inventories. The U.S. EPA conducts independent validation and verification of GHGRP inventories submitted by operators, and errors are flagged for review and/or resubmittal of the GHGRP. The U.S. EPA also has the potential to undertake enforcement action against reporters for violations of reporting requirements under the GHGRP (40 CFR § 98.8).

Completeness [*Low*]: Completeness is low. The GHGRP only includes certain high-emitting facilities, which excludes methane emissions from smaller oil and gas operations that are important

---

[4] ANNEX 3 Methodological Descriptions for Additional Source or Sink Categories, Sections 3.5 and 3.6. https://www.epa.gov/system/files/documents/2022-04/us-ghg-inventory-2022-annex-3-additional-source-or-sink-categories-part-a.pdf.

[5] https://www.epa.gov/sites/default/files/2015-07/documents/ghgrp_inventory_emissions_comparison.pdf

for a full accounting of emissions. The gap in reconciling the U.S. EPA inventory methane emissions versus observed estimates in the oil and gas sector described above is likely due to unaccounted emissions sources.

Inclusivity [*Medium*]: Inclusivity is ranked as medium. The work is performed by technical experts but the GHGI includes a 30-day public review process in which feedback from the public may be incorporated.

Communication [*High*]: Communication is high. The information is published on a public website, with explanatory information and summary statistics and visualizations targeted at non-expert audiences.

4.4.2 The Permian Methane Analysis Project

PermianMAP is a research initiative of EDF and other research partners employing a variety of multiscale technologies including aircraft and satellites to measure methane emissions over a 10,000 km^2 area in Texas and New Mexico, covering over 11,000 oil and gas wells operated by a myriad of operators. The project began collecting data in autumn 2019 and data collection is ongoing. The detailed methodology and data are publicly available.[6] Multiple peer-reviewed papers have employed these data (Lyon et al., 2021; Shen et al., 2022; Zhang et al., 2020) and concluded that the methane emissions in the Permian Basin are higher than activity-based methods such as GHGI (Case Study 4.4.1). In addition, the PermianMAP project includes an operator dashboard[7] that provides measured emissions during the study period after it has been verified by the various operators and other stakeholders.

In the evaluation below, we consider GHG emissions information from the PermianMAP website and users of the information including national decision makers, researchers, and the public.

Usability and timeliness [*High*]: The data are specific to individual facilities and can be used by operators to assess performance at the time of measurement. Information is reported shortly after collection, so timeliness is high after each campaign. The data and results can be used in comparison to activity-based inventories such as GHGI or GHGRP.

Information transparency [*High*]: PermianMAP scores high on transparency because methods and the data are freely available from the project website.

Evaluation and validation [*High*]: Evaluation and validation is high. Different types of measurements are employed and compared, and there is independent evaluation and verification from academic experts.

Completeness [*Medium*]: Completeness is medium. The GHG emissions information produced is only for methane sources from oil and gas facilities in the Permian Basin, but PermianMAP has clearly defined boundaries and covered all oil and gas sources within the boundary. The facility-level analysis is based on spot measurements (typically a few measurements per emission source), so it does not reflect daily variations of emissions or that super-emitters are stochastic.

Inclusivity [*Low*]: Inclusivity is low because local stakeholders were not part of the project design. The project does have a mechanism to communicate with well operators and collect feedback on emissions and responses by operators, though facility-level measurements have stopped as of November 2021, hindering timely responses. Inclusivity could be improved by engaging local communities in this effort.

Communication [*Medium*]: Communication is medium. A detailed, interactive website presents maps, quantitative data, trends, and more. Information targeted at well operators is generated

---

[6] https://permianmap.org/
[7] https://data.permianmap.org/pages/operators

Case Study	Usability and Timeliness	Information Transparency	Evaluation and Validation	Completeness	Inclusivity	Communication
EPA GHGRP and GHGI	Medium	High	Medium	Low	Medium	High
PermianMAP	High	High	High	Medium	Low	Medium

Pillars

FIGURE 4-5 Qualitative evaluation of the U.S. Environmental Protection Agency (EPA) national greenhouse gas inventory (GHGI), Greenhouse Gas Reporting Program (GHGRP), and the Environmental Defense Fund's Permian Methane Analysis Project (PermianMAP) against the six pillars.

for and communicated with them. The methodology is documented on the website, but there is a lack of context and no background information for the general public.

In comparing the evaluation of the U.S. EPA GHGRP and GHGI and the PermianMAP study of oil and gas methane emissions in the Permian Basin, it is evident how different types of GHG information have different attributes and can complement each other (Figure 4-5). By using information from the two products together, decision makers have a better view of the actual emissions in the region and the issues that underpin the mitigation of methane emissions of this type in the Permian Basin, some of which can apply to other regions. Both approaches could improve inclusivity, particularly by engaging with local communities. Improving completeness in both approaches would also result in better information, and PermianMAP could link to and build on public-targeted information from the U.S. EPA to improve communication.

4.5 VERIFY

VERIFY[8] is an EU-funded project that aimed to develop a system primarily for providing science-based support to EU countries' emissions reporting to UNFCCC (Janssens-Maenhout et al., 2020). Part of the project included an assessment of the user requirements and needs from the GHG monitoring, reporting, and verification perspective, as well as challenges and potential for future science (e.g., quantifying different gases, spatial/temporal scales, observation resources, etc.). VERIFY identified three major topical research areas for developing verification tools: (1) fossil fuel CO_2 emissions, (2) terrestrial CO_2 sources and sinks and carbon stocks, and (3) methane and N_2O emissions (see Figure 4-6).

VERIFY integrated atmospheric measurements, satellite observations, and ecosystem data with emissions inventories via modeling and data assimilation to produce country-level GHG budgets for major GHG gases (CO_2, CH_4, and N_2O) for the EU domain using a combination of activity-based and atmospheric-based approaches. VERIFY explored the use of atmospheric tracer or co-emitted species measurements, such as ^{14}C, CO, and NO_2, to estimate fossil fuel CO_2 emissions by separating them from other sources and natural sinks. This approach required improvements to their inventory approach and emission models for co-emitted species and data assimilation (Petrescu et al., 2021; Super et al., 2020). Other GHG emission models, including ecosystem models and land surface models, were improved to provide non-energy CO_2 and methane and N_2O process-specific emissions information more accurately (e.g., Balkovič et al., 2020; Schelhaas et al., 2018; Susiluoto et al., 2018). The activity-based emission models and atmospheric-based approaches incorporating them were used to produce GHG budgets with uncertainty assessments primarily for EU member countries (e.g., Petrescu et al., 2020, 2021).

---

[8] https://verify.lsce.ipsl.fr/

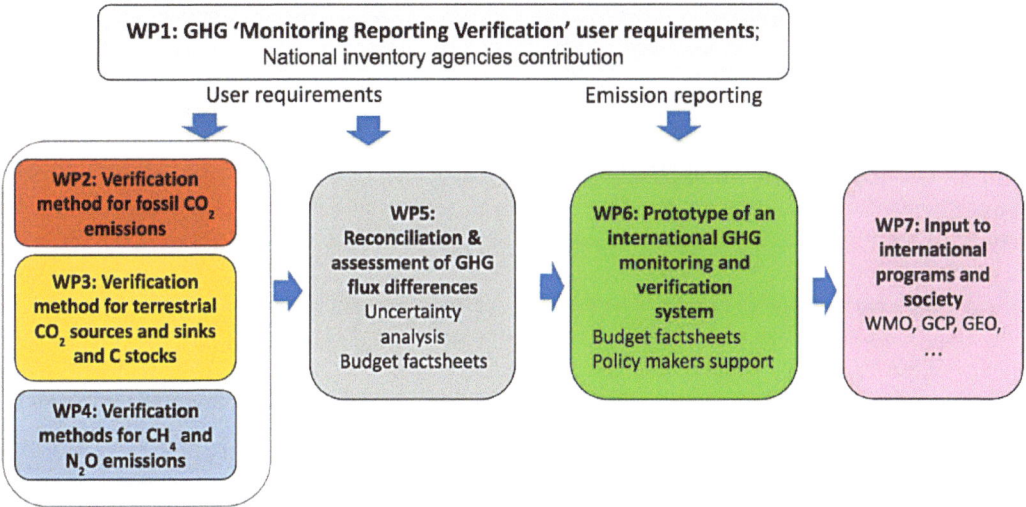

FIGURE 4-6 Schematic overview of the VERIFY structure. SOURCE: https://verify.lsce.ipsl.fr/index.php/presentation/approach.

While an EU-led and EU-focused project, VERIFY also considered how such efforts could be implemented in an international framework for GHG verification support to countries and regions outside of the EU domain. VERIFY also invested in regional and large-scale carbon cycling (e.g., river fluxes, Gommet et al., 2022), including oceans (Roobaert et al., 2018), as well as climate policy analysis (Perugini et al., 2021; Peters et al., 2020), to support the global collective assessment of climate mitigation (e.g., Global Stocktake).

In the evaluation below (see also Figure 4-7), we consider EU-wide and country-level GHG emissions information available from the VERIFY website and academic publications, and users of the information to include national decision makers, researchers, and the public in the European Union.

Usability and timeliness [*Medium*]: Usability and timeliness is medium. The project aims to serve the policy maker community, and they have also had conversations with relevant international organizations (e.g., UNFCCC, the World Meteorological Organization, the Group on Earth Observations), indicating some relevance to stakeholders. While data and reports are available, there has been little evidence that VERIFY products have had an impact on decisions and reported inventories so far. Efforts initiated in VERIFY are likely to continue and become more useful with further development.

Case Study	Pillars					
	Usability and Timeliness	Information Transparency	Evaluation and Validation	Completeness	Inclusivity	Communication
VERIFY	Medium	High	High	High	Medium	Medium

FIGURE 4-7 Qualitative evaluation of VERIFY against the six pillars.

Information transparency [*High*]: Transparency ranks high. The GHG budgets and estimates produced by VERIFY were presented as science synthesis reports[9] and the results, including the budget data sheets and data, are available from their website and data repository for public use.[10] Additionally, the data and plots are available from their interactive tools for improved user accessibility.[11] VERIFY also make some model codes available via GIT.[12]

Evaluation and validation [*High*]: VERIFY ranks high on evaluation and validation from a scientific standard because there have been efforts to address uncertainties and to apply and compare different approaches. Studies evaluating the results are published in reports and peer-reviewed literature.

Completeness [*High*]: Completeness is high. VERIFY provides emissions estimates of sources and sinks for CO_2, methane, and N_2O for the European Union. The spatial coverage is limited to the European Union, as was intentional in the project design. F-gases were also analyzed in the project, and ocean and climate projection components were also included. Reliable atmospheric-based estimation of fossil fuel CO_2 is pending further development.

Inclusivity [*Medium*]: Inclusivity is medium because VERIFY is primarily an EU project with some non-EU collaborators, nearly all academic researchers. The project intended to prototype an international system, which could improve its inclusivity. Inclusivity would be improved by involving all potential users to define user requirements and making efforts to gather public comment.

Communication [*Medium*]: Communication is medium. Their website provides information on project aims, key reports, and some summary graphics of findings. The VERIFY tools and datasets are not always in easily usable or in downloadable formats for a wide range of users.

4.6 Climate TRACE

Climate TRACE (Tracking Real-time Atmospheric Carbon Emissions) is a hybrid approach that utilizes nontraditional data including satellite imagery with artificial intelligence and other data science tools to develop an activity-based account of global GHG emissions. Climate TRACE launched in 2020 and released its first data product in 2021. With 12 member organizations and more than 50 contributors—including academic researchers, nonprofits, tech companies, and climate leaders—it accesses high-resolution satellite imagery and develops methodological approaches to detect and estimate GHG-emitting activities and emissions. Utilizing data from more than 300 satellites, over 11,000 sensors, and other emission sources (Climate TRACE, n.d.), Climate TRACE aims to evaluate GHG-emitting activities in 10 sectors, including agriculture, buildings, manufacturing, maritime, mineral extraction, forestry and land use, oil and gas, power, transport, and waste. Their first data release in 2021 reported national emissions totals in CO_2-equivalents by subsector for the year 2020. They have also engaged with nations and subnational entities to obtain data. For example, Dalmia Cement, one of the largest cement companies in India and a core partner of Climate TRACE, has provided all of its facility-level emissions data to train the platform's estimation of manufacturing-related emissions.

Due to the proprietary nature of some of the data sources underlying Climate TRACE's emissions models and estimates, access to underlying data and source code is not available to the public. Only one and three subsectors out of 14 have code and data sources available, respectively. Methodology documents are limited to a few sectors generally detailing how sectoral emission estimates are derived, including information regarding data sources used, such as satellite data and ground-truth data. Since different organizations lead different sectoral emissions estimates, the

---

[9] https://verify.lsce.ipsl.fr/index.php/repository/key-reports-verify
[10] https://verify.lsce.ipsl.fr/index.php/products
[11] https://verify.lsce.ipsl.fr/index.php/products?view=article&id=153
[12] https://git.nilu.no/VERIFY/CIF

level of methodological granularity and quality varies by sector. For example, the methodology document for power sector emissions modeling prepared by WattTime provides information on the training data, including locations of power plants and remote sensing data to identify these plants' operational status and the machine learning models utilized. Other sectors provide fewer details, such as the aviation sector, which largely refers to the International Civil Aviation Organization's methodology to estimate carbon emissions for air travel and outlines broadly the estimation of aircraft emissions using historical information on flight distances traveled.

In the evaluation below (see also Figure 4-8), we consider GHG emissions information publicly available on the Climate TRACE website and users of the information including national decision makers, researchers, and the public.

Usability and timeliness [*Low*]: Usability and timeliness is currently low because there is little evidence that their products have been utilized by decision makers to date. Climate TRACE provides national-level statistics, which can have limited utility for mitigation decisions. They released country-specific reports in the lead up to COP-26 (26th session of the Conference of the Parties). There is a framework for providing timely updates to data, but in most cases, only data for 2020 are currently available. The integration of data from a range of sources and more real-time information updates have the potential for high usability, but that has not been realized currently. Providing information at subnational scales may expand the number of potential users.

Information transparency [*Low*]: Transparency is currently low. While there is methodology documentation publicly available on their website, transparency is limited by the proprietary nature of some of the data sources underlying Climate TRACE's emissions models and estimates. Access to underlying data and source code is not currently available to the public.

Evaluation and validation [*Low*]: Evaluation and validation is also currently low. Some of the underlying models have been published in the peer-review literature, and some data have been published as part of other thoroughly evaluated projects such as EDGAR. Currently, Climate TRACE-specific reports are only in the gray literature and there have been no peer-reviewed publications about their data product or any associated published evaluation against independent data.

Completeness [*Medium*]: Completeness is medium. Strengths in completeness come from the large number of data sources and sectors included. However, emissions are reported in CO_2 equivalent and, while this metric may be useful for reporting comparable values to the general public, information on individual GHGs is not currently available. Quantifying each GHG is important for mitigation decisions and would support information usability and evaluation and validation against other datasets. The data currently quantify emissions by country, globally, but do not provide spatially disaggregated information.

Inclusivity [*Medium*]: Inclusivity is medium. The project is conducted by a large team of scientific and private-sector partners and, as described above, Climate TRACE has collected some information through engagement and partnership with local companies. Inclusivity could be strengthened through engagement with other types of local stakeholders (e.g., air quality agencies) or the incorporation of citizen science.

Case Study	Pillars					
	Usability and Timeliness	Information Transparency	Evaluation and Validation	Completeness	Inclusivity	Communication
Climate TRACE	Low	Low	Low	Medium	Medium	Medium

FIGURE 4-8 Qualitative evaluation of Climate TRACE (Tracking Real-time Atmospheric Carbon Emissions) against the six pillars.

Communication [*Medium*]: Communication is medium. An interactive website allows for data exploration and the data are available to download in usable formats (csv, json). Communication could be improved with online tutorials and more information on the webpage explaining the data and methods. Detailed documentation is not available in the gray literature reports that are publicly available. The metric of CO_2 equivalent is difficult for decision makers to interpret and for researchers to make comparisons to.

Themes of Framework Evaluation for General Approaches and Case Studies

This chapter outlined six pillars for evaluating GHG emissions information, provided a qualitative evaluation of the capabilities of current approaches to satisfy these pillars, and presented a set of case studies to demonstrate how this framework could be applied to current efforts. While the case studies and approach rankings were not exhaustive of all possible applications, a number of themes emerged that have informed the recommendations the Committee makes in the following chapter.

Overall, usability and timeliness ratings are often low because the information is not available at relevant time scales and at the level of granularity needed to be useful; complex methods also tend to remain within a small, technical community. Information transparency tends to rate more highly because the information is often published and publicly available. Risks to information transparency include information behind paywalls, private data collection, and proprietary methodologies. While some form of evaluation and validation is often performed, there are opportunities to improve, particularly for activity-based data, and move beyond internal consistency checks to evaluate information against independent data and consistently develop and report uncertainty estimates.

For completeness, a primary limitation is that all GHGs are often not included, especially in atmospheric-based approaches. Areas for improvement in activity-based approaches include capturing all sources and sinks, representing nonroutine emissions, and applying appropriate emission factors. Inclusivity is low in most cases, particularly when considering historically underrepresented communities and geographies. Inclusivity could be enhanced by collaborating with stakeholders to develop methods and priorities, training and building the capacity of underrepresented communities, and using citizen scientists to make observations. Finally, communication is low in many cases when GHG emissions information is developed for technical communities and not adequately communicated to policy makers and the public. On the whole, GHG data are scattered across a large number of portals and repositories. However, some projects have shown promise by creating websites with easy-to-navigate information, explanatory background, and data in formats that are easy to use.

The case studies demonstrate how the framework of six pillars for evaluating GHG emissions information could be applied more specifically to current efforts to identify their strengths and the areas where improvement or supplemental information is needed. Within a given approach, there can still be a wide range of performance relative to the pillars depending on the application, the user, and the user's needs. Case Study 4.4 on methane emissions in the Permian Basin showed how different types of GHG emissions information can be combined to produce a better picture of actual emissions. Efforts to move toward explicit hybridization of different approaches and datasets holds promise for combining the strengths of activity- and atmospheric-based approaches (INFLUX, VERIFY) and for mining untapped data sources (Climate TRACE), although these efforts may remain unused by decision makers if they do not target the needs of decision makers with transparent methods and data that are communicated well. As demonstrated again by the case studies, inclusivity of GHG emissions information is an area that needs improvement if the best information is to be produced.

5

Recommendations

Three converging trends motivated this report: (1) rapidly increasing demand from a range of users for trusted information about greenhouse gas (GHG) emissions across multiple sectors and geographic scales; (2) development of many new approaches for quantifying GHG emissions information that aim to address this increasing demand; and (3) a growing and rapidly evolving institutional landscape, including public, private, and academic entities seeking to provide better GHG emissions information. As described in the previous chapters, these trends have cultivated an exciting moment of innovation and improved capabilities. At the same time, this evolving landscape can be complicated and confusing to navigate. In the following recommendations, the Committee strives to find a reasonable balance between more effectively coordinating the large amount of GHG emissions information that currently exists, encouraging the creative energy and urgency to quickly advance new capabilities, and establishing the need for approaches and institutions that will lay the groundwork for trusted and useful information.

Current approaches for quantifying GHG emissions—mostly activity-based approaches that utilize activity data as representative indicators to calculate GHG emissions and atmospheric-based approaches that use measurements of atmospheric concentrations to infer emissions information—have been well developed in the research community and are being used, to some extent, to inform decision making. However, there is an opportunity to improve current approaches to maximize the adoption and usefulness of the information in response to needs from decision makers. For example, some activity-based approaches have been developed and refined for years and thus have well-established protocols that have been repeatedly documented, evaluated, and validated; however, some long-standing data gaps and uncertainties have proven hard to resolve. Although some approaches are well developed, they may not be easily adopted due to various constraints. For example, stakeholders in countries within Latin America, Asia, Africa, and Oceania (excluding Japan, South Korea, Singapore, and Israel) often lack sustained funding to develop and maintain technical capacity for regular GHG emissions information development and reporting.

In this chapter, the Committee introduces recommendations to both enhance current GHG emissions information development capabilities and strive for hybrid approaches that would optimize the integration of individual activity- and atmospheric-based approaches in order to provide the best available GHG emissions information for users. The six pillars introduced in Chapter 4 underpin the recommendations and their applications, described below. The previous chapters have considered the full range of spatial scales relevant to users of GHG emissions information: global, national, regional, local, and facility. The Committee's recommendations are intended to apply to this spectrum of geographic scales with a minimum common focus on the national to global scale. Furthermore, hybrid approaches, detailed below, could have the benefit of integrating GHG emissions information across all scales, improving both the provisioning of GHG emissions information and decision making. Lastly, the committee recognizes the varying capabilities and constraints of different members of the global community and that the different decision-making phases—planning, tracking, and assessment and verification—will have different needs.

Figure 5-1 presents a conceptual overview of how the pillars described in Chapter 4 provide a framework for the evaluation of GHG emissions information and approaches underlying their development, and how this relates to informing different stages of decision making in the context of a policy goal. Decision makers often have different GHG emissions information requirements and needs in each phase. While much of the scientific research in GHG emissions information focuses on method development for assessment and verification, many decision makers across scales are in the planning phase. Decision makers are assessing what they *need* to do, rather than what they have done. In this phase, they need to understand the GHG emissions information available to them so they can adequately plan and execute appropriate climate mitigation actions. The pillars underlie the knowledge-to-action circle depicted in Figure 5-1 in which the development of GHG emissions information would strive to satisfy the pillars in order to be actionable by decision makers in support of their policy goal. Furthermore, different pillars may be more important during different decision-making phases. Thus, the framework allows decisions makers to consider what phase they are in and which pillars will best suit their needs.

In general, inclusivity and equity remain an issue to decision makers at all spatial scales. Although low-income countries and communities share minimal responsibility for current and his-

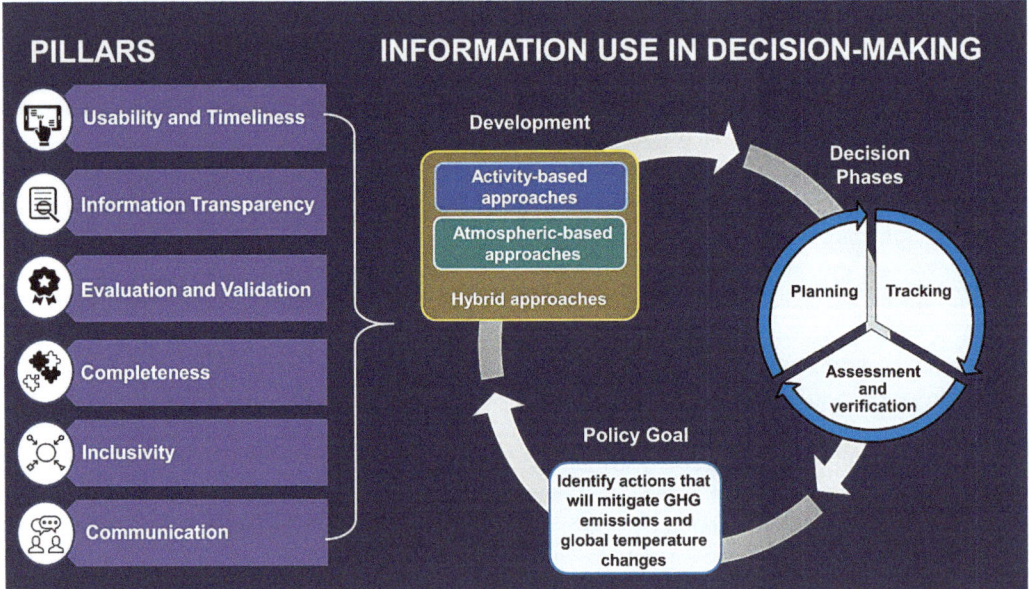

FIGURE 5-1 Conceptual description of framework for evaluating greenhouse gas (GHG) emissions information. The six pillars (left) are used to drive the iterative phases for the development and use of information in decision making (right).

toric anthropogenic GHG emissions, in general, the impacts of climate change disproportionately affect countries within Latin America, Asia, Africa, and Oceania (excluding Japan, South Korea, Singapore, and Israel). Thus, these countries must often prioritize climate adaptation rather than mitigation, including the quantification and attribution of GHG emissions. In building a globally coordinated GHG emissions information system, attention and investments to engage the global community are vital. Centrally, implementation of the recommendations that follow should include participation, communication, and knowledge exchange among scientists, users, and decision makers globally, particularly from Latin America, Asia, Africa, and Oceania. While these recommendations include specific points about enhancing global inclusivity in the development of GHG emissions information, ultimately, the needs and priorities for adapting to and mitigating climate change must be determined by countries and regions themselves.

The Committee expects that these recommendations will be implemented by a wide diversity of actors in the global community: local municipalities to national-scale governments with varying capacities, the United Nations Framework Convention on Climate Change (UNFCCC) and other international bodies, individual companies seeking to understand their own emissions, and large philanthropies and nongovernmental organizations seeking to develop crosscutting products. With this in mind, recommendations are offered that set clear directions that would be relevant across multiple audiences, and empower different entities to identify specific implementation steps within their means and to meet their needs.

Advancing Greenhouse Gas Emissions Information Capabilities, Trust, and Accessibility

Greenhouse gas emissions information development and evaluation should strive to align with the six pillars: usability and timeliness, information transparency, evaluation and validation, completeness, inclusivity, and communication.

The "pillars" presented in Chapter 4—usability and timeliness, information transparency, evaluation and validation, completeness, inclusivity, and communication—provide a way to evaluate individual emissions datasets and approaches, as illustrated through the case studies. Not surprisingly, there is a lot of variation among different approaches in meeting these criteria. More established protocols—for example, emissions reporting as part of the UNFCCC process—have put in place some standards around transparency and evaluation. New innovative ways are continually being identified to address challenges with data access or uncertainty quantification, but most of these approaches have not yet been rigorously evaluated. Thus, the pillars provide guidance for improving GHG emissions information development and products.

These same pillars embody the desired attributes for the institutions that develop GHG emissions information and the broader aspirations for the global, multiscale endeavor of understanding the sources and sinks of GHGs. Strengthening GHG emissions information to satisfy these pillars at the local and subnational level would help to build international coordination and support. A good practice would be for GHG emissions information providers to self-assess their adherence to the pillars. Such self-assessment would be particularly valuable in cases where multiple sources of GHG emissions information developed for different intended purposes are being used or integrated into a different product with a different intended use. The application of the six pillars of the framework to both individual datasets and approaches as well as the structures that support the development, provision, and exchange of GHG emissions information would advance the current complex GHG emissions information landscape toward one that could more comprehensively meet the needs of users and decision makers.

Greenhouse gas emissions information should be better coordinated (e.g., through the creation of a coordinated repository or federation of repositories) across the global community, enabling adherence to a set of minimum common pillar attributes.

As discussed in Chapter 2, many different GHG emissions information sources have been developed for both scientific and decision-making needs. Currently, some information can be accessed through individual data portals, but is often provided in different formats and varying levels of completeness, largely driven by the specific purpose initiating the data collection or model estimation effort; in some cases, the information is behind paywalls. As a result, it can be challenging to integrate and compare multiple data sources. A coordinated repository or federation of repositories where GHG emissions information can be hosted, documented, and clearly characterized would be a critical step forward in maximizing use and understanding of GHG emissions data products. A mechanism that brings different types of information together could facilitate the integration of multiple types of data at various spatial scales and make the information accessible in a timely manner to decision makers in ways that meet their needs. Such a clearinghouse or federated data center could establish standards and practices that enable decision makers and other data users to clearly and quickly grasp individual characteristics, comprehensiveness, transparency and traceability to source material, and citations for the wide range of GHG emissions information. It could also facilitate input from decision makers on future GHG emissions information development and specifications. To maximize adoption and equity, such a coordinated effort should support and integrate information from existing national and international programs—for example, the UNFCCC reporting program—to leverage, rather than replace, both established and emerging efforts. Countries, particularly developing countries, have spent time and effort to build the capacity needed to calculate reliable inventories using the IPCC guidelines, and there is an opportunity to build on this existing capacity. As a case in point, the International Methane Emissions Observatory is designed to gather information on methane emissions from multiple sources, including national inventories reported to the UNFCCC, corporate reporting, remote sensing data, and measurements of emissions through scientific studies (UNEP, 2021a).

Critical characteristics and functions of a coordinated repository or clearinghouse would operationalize each of the six pillars by including

- Timely information that is transparent and traceable to primary, supporting, and derived datasets;
- Standardization of data formats and metadata to facilitate comparability and interpretability across scales;
- Descriptive documentation of models and estimation procedures in nontechnical language, and in multiple languages, that can be used to enable the use of diverse information, including novel observations and methods from the research community;
- Qualitative (e.g., caveats and limitations) and/or quantitative (e.g., uncertainties, error characterization) evaluation metrics;
- Databases of key input data and information (e.g., emission factors, activity data, atmospheric observations, models) that would be regularly updated and widely accessible to facilitate information exchange;
- Governance mechanisms that are coordinated, trusted, and designed to be inclusive of the global community and built on the best practices of data governance and information quality;
- Education modules and capacity building for using GHG emissions information and contributing data and estimation results; and

- Mechanisms to support stronger collaborations between GHG, air quality, and meteorological science communities and stakeholders.

Greenhouse gas emissions information providers should clearly communicate underlying data, methods, and associated uncertainties.

While the information clearinghouse or federated repository effort described above would be a longer-term undertaking for the global community, actionable steps can be taken by data providers in the short term to enhance the transparency of GHG emissions information where feasible. Focused resource allocation or government purchases aimed at bringing data and methods into the public domain with standards on transparency and open principles (e.g., FAIR data principles [findable, accessible, interoperable, and reusable]) could have substantial near-term impacts on the utility of GHG emissions information. Following many of the same guidelines outlined for a clearinghouse and aligning with the pillars, data providers have the opportunity to facilitate comparability and verification of their data and methods to foster trust between information providers and users. In addition, communication is an important consideration in the presentation of datasets to potential users.

As the international community more seriously takes up the need to reduce GHG emissions, decisions based on GHG emissions information will have significant implications for governmental policy making and regulation, business and financial outcomes, and community and household planning. Arguments about the validity of data sources could lead to delay in action and inaccurate information could lead to costly errors in determining how to mitigate emissions. To foster trust in GHG emissions information, transparency of the process and methods of data collection, transformation, analysis, evaluation, and validation is of utmost importance. In particular, transparency is essential for knowledge and resource sharing in the global community, and, specifically, to build capacity in regions with less GHG emissions information development capabilities (e.g., the Global South).

Many atmospheric- and activity-based GHG emissions information products and research efforts span a range of methodological and data transparency standards. While some approaches are documented in the peer-reviewed literature and many scientific journals encourage or require authors to make the data publicly available, the highly technical language used is complex, limiting the transparency of methodologies for users of GHG emissions information. Many atmospheric observations are publicly available, but the analytical tools necessary to use the data may not be. Modeling codes are not often available, and datasets, including those produced by the private sector, are often behind paywalls and may not include information about the sources of the data.

Addressing Key Data and Information Gaps and Uncertainties

Greenhouse gas emissions information (e.g., observations, data analysis, activity data, emission factors) development at more granular temporal and spatial scales with source-level detail should be accelerated to meet the rapidly increasing needs of cities, states, and provinces for managing their emissions.

Beyond the traditional global decision makers—most commonly identified at the international/national scale—are a large and growing community of decision makers and users that have emerged as critical policy actors in the last two to three decades. Cities, states, provinces, landowners, and the business community, among others, are organizing collectively, enacting mitigation policies, and in critical need of consistent, standardized, trusted GHG emissions information. For example, for governments or the private sector to meet their planned emission mitigation targets, they will need to quantify their baseline GHG emissions, identify the most effective mitigation strategies, and track progress toward meeting their targets.

Where possible, there are substantial gains associated with extending the GHG emissions information development identified thus far in this report to these finer space and time scales, enhancing completeness. Not only does this dramatically increase the user base for the GHG emissions information but it can also improve the accuracy and robustness of GHG emissions estimation at all scales. Furthermore, enhancing source-level detail—for example by characterizing the entire distribution of emission sources—would strengthen completeness of GHG emissions information and better inform mitigation decisions. This is particularly important for methane that has a mixture of point and nonpoint sources, with varying degrees of temporal and spatial emission distributions. Enhancing GHG emissions information itself and communication among decision makers at multiple scales could have a considerable policy benefit in countries with multiscale governance systems. Finally, information at finer space and time scales has the potential to improve methods and pilot new observational capabilities that could be scaled up. Information about GHGs at these more granular spatial scales may also enhance efforts to evaluate and validate national and global scale information. The clearinghouse or federated repository recommended by the committee would be intended to bring information across spatial scales together in a central location or coordinated set of locations.

Currently, data available on granular spatial and temporal scales are insufficient and there is a need to expand the necessary data resources. This includes activity data, emission factors, and atmospheric observations. Methods catered to smaller scales that utilize various datasets to generate GHG emissions information also need to be developed and best practices established. Pilot projects including atmospheric observations currently running at city scales could be extended and expanded to other locations, with particular attention paid to inclusivity and usability to maximize effectiveness and impact.

The accuracy and representativeness of all underlying data used to estimate greenhouse gas emissions (e.g., emission factors, activity data) should be further improved.

Many of the data elements, observations, and models used to estimate GHG emissions rely, sometimes by necessity, on large spatial averages or averages that represent well-observed or high-capacity parts of the globe. The Committee recognizes the need to improve both the specific representativeness and resolution across the globe of these key underlying data drivers to strengthen the completeness and accuracy of GHG emissions information. Examples include emission factors that are often fuel averages (i.e., missing true coal quality variation or biomass variation) or tied to countries with well-quantified fuel characteristics; activity data collected for particular countries but used to calculate emissions for other countries even if the data are unrepresentative; and atmospheric monitoring that may only reflect large-scale integration of information or is biased to locations with high scientific capacity.

Similarly, while emission factors for fossil fuel combustion CO_2 emissions are better known due to more predictable fuel sources and infrastructure types, emission factors for non-energy-sector emissions (e.g., agriculture, forestry, and other land use and waste) or non-CO_2 GHGs such as methane are less well known due to the heterogeneity of the sources and emissions processes. Additionally, emission factors are typically determined for countries in the Global North and may represent averages over large spatial scales (e.g., national, regional), and thus are not representative of locations in the Global South and specific local conditions more broadly. Although activity data may be a larger source of uncertainty for national and subnational activity-based emissions estimates, particularly in the Global South, unrepresentative emission factors can also result in large uncertainties and errors in emissions inventories developed using activity-based approaches.

Finally, widely used emission factors from databases are often out of date and not regularly updated with new information, and the use of generic emission factors in activity-based approaches can lead to overestimates of emissions. Countries and localities that do not have the resources to invest in updating emission factors specific to their local emission sources are relegated to using outdated information that is biased toward Global North[1] countries. On the other hand, the use of more refined emission factors can lead to underestimates in emissions estimates by not accounting for abnormal conditions. The lack of dynamic updates to emission factor information may also limit opportunities to take advantage of novel approaches being developed. The current emission factor workflow and infrastructure has not facilitated the use of the best available information to date, and there is an opportunity to devote resources in a way that would improve the quality of emissions estimates. For example, at the facility scale, emissions estimates could be improved by employing supplier-specific data rather than employing national- or regional-level data.

Operationalizing Current Capabilities

Greenhouse gas emissions estimation research efforts should transition with urgency to operational capabilities with institutions to maintain and ensure longevity.

As the urgency to immediately reduce GHG emissions is increasing, decision makers likewise need the best-available information about emissions as soon as possible. The typical pace of research and development for new approaches, and for bringing approaches from research into operations, is too slow to meet this demand. Accelerating the transition of research to operations will require scientists, research funders, and data users to identify ways to lower existing barriers to that transition and ways to make new data products more immediately usable. The clearinghouse or federated repository recommended above, along with alignment with the pillars, should help make new GHG emissions information usable more quickly.

Some of the approaches for generating GHG emissions information that have been developed within the scientific community are, or are approaching, a level of readiness that would allow for operationalization, resulting in consistent, reliable information that could be better embraced and exploited by decision makers. New highly granular activity-based approaches and much of the work on atmospheric-based approaches to date have largely been limited to the expert scientific community and have not yet been fully embraced in the decision-making process, even though their potential has been recognized (e.g., IPCC, 2019a).

Some countries already use atmospheric-based approaches to assess their non-CO_2 GHG emissions annually, and the Copernicus Atmosphere Monitoring Service and Global Carbon Project use atmospheric-based approaches to assess global non-fossil fuel CO_2 fluxes annually. Limitations to incorporating the atmospheric-based component of estimating fossil fuel CO_2 emissions and other gases to the operational stage on a global scale are the lack of sufficient observations and a specific infrastructure at one or more institutions that perform the calculations operationally.

Advanced activity-based research efforts have remained in the research realm but similarly show promise within operational systems. Many of these new efforts are availing of new data, sometimes specific to high-income countries, but there is also use of globally available, though potentially costly, remote sensing data. Barriers to operational application include efforts to collect more activity data in countries and regions where limited research into advanced techniques has

---

[1] Note: The IPCC Emission Factor Database was created using data based on OECD (Organisation for Economic Co-operation and Development) countries.

occurred. Enhancing inclusivity by enlisting stakeholders to help develop methods and priorities, training and building capacity of underrepresented communities, using citizen scientists to make observations, and leveraging local expertise to design and optimize atmospheric measurement networks would improve atmospheric-based approaches. Furthermore, infrastructure on high-volume data ingestion and processing is essential. Finally, increased granularity is needed in much of the globe in order to achieve data completeness, which, in turn, places emphasis on both ground-based and remotely-sensed information on emitting infrastructure with greater coverage.

Much like weather forecasting systems, optimal approaches will combine a more complex suite of observed data to drive optimal estimates of GHG emissions. For atmospheric-based approaches, this will combine ground- and aircraft-based observations that have high accuracy and sensitivity to surface emissions with satellite observations that have global spatial coverage. A key element may include the measurements of tracers that can attribute specific emission sources. Some of these tracers are already monitored by the air quality community, so the collaboration between the GHG and air quality institutions would support this effort. Current atmospheric observations are not sufficiently representative or dense. More representative and higher density of ground- or aircraft-based observations of GHGs and tracers such as isotopic measurements of CO_2 and methane and ancillary gases are needed to move toward useful and well-constrained national-, state-, or city-scale estimation. More observations of atmospheric transport and dynamics—specifically meteorological parameters and boundary layer height—in the low atmosphere are also needed and would be aided by stronger collaboration with meteorological institutions. Satellite observations also need improvement in resolution, accuracy, and precision and reduced sensitivity to clouds, aerosols, and land surface characteristics. Iteration between scientists and decision makers is needed to prioritize the continuity and deployment of observing systems and to ensure these approaches are integrated with the decision-making process.

Striving for Hybrid Approaches

Greenhouse gas emissions data collection, modeling, and information development should be designed and implemented to enable a fuller integration and "hybridization" of information and approaches.

Most of the current GHG inventory and information development to date has tended to use single methods or approaches with single-technique observations or data sources (e.g., CO_2 mixing ratios, energy consumption data). While this approach was warranted during the development of many of the state-of-the-art estimation techniques, going forward, a "cross-technique" or hybridization of (traditional) approaches and datasets would provide more accurate GHG emissions information by integrating different types of information with more granularity. Some of this work has begun and includes new machine learning and other nonparametric numerical techniques that leverage new data from private and public satellites, sensors, and other types of activities.

Efforts that more fully integrate the traditional activity- and atmospheric-based approaches also present a path toward a more integrated, complementary approach that overcomes gaps and weaknesses in each approach when used in isolation. For example, deeper integration of atmospheric transport models and emissions models or algorithms allow for more direct use of hybrid approaches that need further development and support. This approach is more akin to data assimilation techniques, which have matured in the numerical weather prediction community, among other fields. A single dynamical model can then directly use all observations at higher fidelity—for example, using atmospheric mixing ratios, observed traffic data, and fuel consumption data directly to produce the best estimate of fluxes, calibrated to the uncertainty associated with each observed quantity—is another example of such a hybrid approach. Greater synergy between air quality, meteorology, and GHG emissions information data collection and analysis efforts would facilitate the development of these hybrid approaches.

Recent developments in multiple dataset integration through advanced ML and related numerical schemes can uncover new statistical relationships and patterns between predictors and GHG emissions data. These techniques also offer opportunities to integrate a wider and more diverse set of data relevant to GHG emissions and fluxes with activity- or source-based inventories. Enabling the development of greater integration and hybrid approaches would require designing data collection to fill the most needed gaps. To strive for hybridization is to holistically improve GHG monitoring across scales, approaches, and capacity. Hybrid approaches that mix complementary approaches, datasets, and models and that integrate the needs of end users promise to offer richer, more usable data outputs for decision making.

Ensuring Usability, Timeliness, and Effective Communication of Greenhouse Gas Emissions Information

Greenhouse gas emissions information generators, decision makers, and global stakeholders should engage in an iterative process in a timely manner to ensure the information provided is relevant and useful.

Incorporating decision-maker input is critical for information developed to respond to the policy needs of stakeholders and decision makers. The time lag to integrate relevant findings from new research into developing empirical- or measurement-informed inventories limits development and execution of sound mitigation policy, and delays transmittal of appropriate market signals for investments and technology development related to mitigation programs by various stakeholders. Usability and timeliness of GHG emissions information can be enhanced if data producers and users engage in an iterative process, which the clearinghouse or federated repository could support, to facilitate investments in systems that are focused on providing decision support and responsive to an evolving policy-making landscape.

The primary motivation for improving the quantification of GHG emissions is to guide decisions—across all scales—that can contribute to rapidly decreasing global emissions. Thus, information needs to be made available in ways that decision makers can readily use. This involves both what kind of information is developed and how the information is presented and described. Improved communication between scientists and researchers working on improving GHG emissions information for end users would enhance communication of results, guide updates to analytical tools to better address users' needs, and inform priorities for new investments in measurements and analyses that align with users' needs. Particular attention is needed to the communication and usability of newer atmospheric-based and hybrid approaches. While these approaches have demonstrated powerful examples of source attribution at local and regional scales, as well as the potential to evaluate and improve activity-based emissions information, they have had more limited utility in the emissions reporting and decision-making context.

As more users are utilizing and communicating information about GHG emissions, and as more data sources become available, clear expectations about how to evaluate information will help build literacy and shared understanding. Drawing on the pillars in the Committee's framework, Box 5-1 outlines key questions that users and communicators of GHG emissions information should pose as they consider how to interpret and use the data.

> **BOX 5-1**
> **Guidance for Evaluating Greenhouse Gas Emissions Information**
>
> The following questions, guided by the pillars, can provide a useful starting point for decision makers, media, and other stakeholders who are evaluating the credibility and usefulness of new greenhouse gas (GHG) emissions information.
>
> - ***Are underlying data, methods, and uncertainties clearly communicated?*** Specifically, information should be provided transparently in formats that are widely used and traceable. It should be made available on websites that are easy to navigate, where data are free to download, with nontechnical documentation, in multiple languages, and in formats that are easy to understand. Standardization of the information—for example, through the guidelines discussed in the clearinghouse above—allows for comparability among GHGs, sectors, time periods, and geographies.
> - ***Over what time period, spatial domain, and for which sources was the information collected?*** GHG emissions information collected for a limited time or in a small geographic area may not be sufficient to draw broad conclusions about an entire emissions sector or geography. Similarly, GHG emissions information collected for a limited number of source types may not be sufficient to draw conclusions about emissions of a GHG more broadly. Data providers should exercise caution when communicating with the public about the results of a single study and clearly explain the limitations of their data and methods. If extrapolations are made in order to complete a GHG inventory or dataset—for example, from a limited measurement campaign to larger spatial and temporal scales—these assumptions should be clearly documented, communicated, traceable, and note which sources were excluded from an analysis.
> - ***Have the approaches and data been appropriately evaluated and validated?*** Evaluation against independent data and methods is key to giving users confidence in GHG emissions information. Data should be assessed for reliability, accuracy, completeness, and quality for the user's intended purpose. Testing the reproducibility of the approach and comparing to independent information are valuable forms of validation. Quantifying and documenting uncertainties to the extent possible in a quantitative or qualitative way would support data evaluation. Ideally, biases and uncertainties associated with the information would be quantified and an assessment of the limitations of the approach would be provided.
> - ***Are multiple data sources or approaches used to support conclusions about greenhouse gas emissions?*** Multiple independent data sources or approaches may support a central conclusion, or help identify gaps or discrepancies where further attention is needed. The strengths of the hybrid approach recommended above come from the integration of multiple approaches and data sources to support completeness, evaluation, and, in some cases, validation of GHG emissions information.
> - ***Have the approaches and resulting data involved locally-based researchers and benefited from stakeholder input and expert review?*** Building local capacity for measurement and monitoring contributes to sustainability and inclusivity. Meaningful engagement with affected populations also increases the inclusivity of GHG emissions information. Increasing access; deploying appropriate technologies and approaches; and clearly communicating data, methods, and uncertainties, for example through a web user interface, are necessary but not sufficient for communities and GHG emissions information users to provide input. Many avenues of engagement, through for example public comment periods, stakeholders workshops, or citizen science data collection opportunities, would enhance GHG emissions information quality and usability.

Concluding Thoughts

The focus of this report has been on analyses and inventories of the anthropogenic emissions affecting atmospheric GHG gases (broadly examined here in terms of all gases and particles affecting the changing radiative forcing on climate). Existing development of GHG inventories has provided significant insights into human-related emissions across all scales. However, as highlighted in the report, there remain major uncertainties in current inventories that limit their use in meeting the extensive, and growing, needs of decision makers going forward. This report highlights a framework for addressing some of these needs, while also mentioning additional issues requiring further consideration. The report will conclude by highlighting some of these issues.

Changes in natural sources—such as those associated with wetlands, permafrost thaw, and wildfires and their associated emissions—can also affect the atmospheric concentrations of CO_2, methane, and other radiatively important gases and particles. In addition, removal processes, or sinks, affect atmospheric concentrations of GHGs. A variety of research studies cited earlier have shown that natural sinks can change spatially and temporally and that they affect the interpretation of emissions inventories. In addition, the changes in climate are themselves affecting natural sources and sinks, with the changing intensity of extreme events being a major factor in these effects. Enhancing understanding of the natural sources and sinks for these GHGs in relation to changing atmospheric concentrations is therefore fundamental to understanding climate change and to the policy responses made by decision makers.

Nonetheless, the integrated overall net emissions have received less focus historically. The focus has largely been on the reduction of fossil fuel emissions, industrially produced (e.g., fluorocarbon) emissions, and emissions from agriculture, land use, and deforestation. Given the importance of land use and land cover change globally, nature-based solutions may be an emerging mitigation and adaptation strategy. A better understanding of the net emissions and associated changes in atmospheric fluxes at all spatial scales is essential for analyzing atmospheric concentration changes for all GHGs and considerations of potential future policy actions. This enhanced understanding will be important to climate plans for all spatial scales, especially with many such planning analyses already beginning to integrate the natural and biogenic components into their inventories; those analyses need more guidance and an enhanced capacity for developing accurate estimates. The global financial sector has an emerging interest and quite possibly new requirements for reporting on GHG emissions and sinks at industrial and business sites and operations.

In addition to the importance of a better understanding of the natural emissions relating to the ocean and unmanaged lands, there is a need for a better understanding of the human-related emissions and sinks associated with managed vegetation. There are especially large uncertainties in net GHG emissions from land use and land-use changes. Improvements in modeling across scales are needed to enhance understanding of emissions and sinks from natural and managed landscapes. Despite an increasing emphasis on capturing these effects in Earth system models, these models still do not capture many important feedbacks and interactions affecting emissions and sinks, limiting understanding of net emissions and the resulting radiative effects.

Overall, significant uncertainties remain in GHG inventories for many parts of the world. Human-related emissions are especially poorly understood for many countries in the Global South. As mentioned earlier, additional observations and analyses are needed to improve emissions inventories for these regions. While low uncertainties may not be needed for every application and decision process, in general, across the planet, reducing uncertainties in the strongest climate forcers (i.e., CO_2, methane, N_2O, and fluorocarbons) is likely important for many policy considerations. A similar emphasis may eventually be needed in reducing uncertainties in inventories for shorter-lived gases and particles. Requirements will likely evolve as decision making evolves.

This report is aimed at analyzing and improving the evaluation of anthropogenic emissions information considered for decision making. The six pillars recommended here provide a framework to evaluate GHG emissions information that can be adapted as GHG emissions information systems become more complex and to serve a range of decision-making needs. In addition, the report includes a number of recommendations toward significantly advancing the accuracy of emissions inventories by better accounting for all available observations and information, while also improving the understanding of the role of human activities in GHG emissions. By examining existing inventories and future needs, the hope is that this report will help push us all forward in assisting the future decision-making process.

References

Agustí-Panareda, A., M. Diamantakis, S. Massart, F. Chevallier, J. Muñoz-Sabater, J. Barré, R. Curcoll, R. Engelen, B. Langerock, R. M. Law, Z. Loh, J. A. Morguí, M. Parrington, V. H. Peuch, M. Ramonet, C. Roehl, A. T. Vermeulen, T. Warneke, and D. Wunch. 2019. Modelling CO_2 weather – why horizontal resolution matters. *Atmospheric Chemistry and Physics* 19(11):7347-7376. https://doi.org/10.5194/acp-19-7347-2019.

Akagi, S. K., I. R. Burling, A. Mendoza, T. J. Johnson, M. Cameron, D. W. T. Griffith, C. Paton-Walsh, D. R. Weise, J. Reardon, and R. J. Yokelson. 2014. Field measurements of trace gases emitted by prescribed fires in southeastern US pine forests using an open-path FTIR system. *Atmospheric Chemistry and Physics* 14(1):199-215. https://doi.org/10.5194/acp-14-199-2014.

Akhshik, M., A. Bilton, J. Tjong, C. V. Singh, O. Faruk, and M. Sain. 2022. Prediction of greenhouse gas emissions reductions via machine learning algorithms: Toward an artificial intelligence-based life cycle assessment for automotive lightweighting. *Sustainable Materials and Technologies* 31:e00370. https://doi.org/https://doi.org/10.1016/j.susmat.2021.e00370.

Al-Shalan, A., D. Lowry, R. E. Fisher, E. G. Nisbet, G. Zazzeri, M. Al-Sarawi, and J. L. France. 2022. Methane emissions in Kuwait: Plume identification, isotopic characterisation and inventory verification. *Atmospheric Environment* 268:118763. https://doi.org/10.1016/j.atmosenv.2021.118763.

Alden, C. B., S. C. Coburn, R. J. Wright, E. Baumann, K. Cossel, E. Perez, E. Hoenig, K. Prasad, I. Coddington, and G. B. Rieker. 2019. Single-blind quantification of natural gas leaks from 1 km distance using frequency combs. *Environmental Science & Technology* 53(5):2908-2917. https://doi.org/10.1021/acs.est.8b06259.

Allen, D. 2016. Attributing atmospheric methane to anthropogenic emission sources. *Accounts of Chemical Research* 49(7):1344-1350. https://doi.org/10.1021/acs.accounts.6b00081.

Allen, D. T., V. M. Torres, J. Thomas, D. W. Sullivan, M. Harrison, A. Hendler, S. C. Herndon, C. E. Kolb, M. P. Fraser, A. D. Hill, B. K. Lamb, J. Miskimins, R. F. Sawyer, and J. H. Seinfeld. 2013. Measurements of methane emissions at natural gas production sites in the United States. *Proceedings of the National Academy of Sciences* 110(44):17768-17773. https://doi.org/10.1073/pnas.1304880110.

Allen, M. R., G. P. Peters, K. P. Shine, C. Azar, P. Balcombe, O. Boucher, M. Cain, P. Ciais, W. Collins, P. M. Forster, D. J. Frame, P. Friedlingstein, C. Fyson, T. Gasser, B. Hare, S. Jenkins, S. P. Hamburg, D. J. A. Johansson, J. Lynch, A. Macey, J. Morfeldt, A. Nauels, I. Ocko, M. Oppenheimer, S. W. Pacala, R. Pierrehumbert, J. Rogelj, M. Schaeffer, C. F. Schleussner, D. Shindell, R. B. Skeie, S. M. Smith, and K. Tanaka. 2022. Indicate separate contributions of long-lived and short-lived greenhouse gases in emission targets. *npj Climate and Atmospheric Science* 5(1):5. https://doi.org/10.1038/s41612-021-00226-2.

Allwood, J. M., V. Bosetti, N. K. Dubash, L. Gómez-Echeverri, and C. v. Stechow. 2014. Glossary. In *Climate Change 2014: Mitigation of Climate Change. Contribution of Working Group III to the Fifth Assessment Report of the Intergovernmental Panel on Climate Change.* O. Edenhofer, R. Pichs-Madruga, Y. Sokona, E. Farahani, S. Kadner, K. Seyboth, A. Adler, I. Baum, S. Brunner, P. Eickemeier, B. Kriemann, J. Savolainen, S. Schlömer, C. v. Stechow, T. Zwickel, and J. C. Minx, eds. Cambridge, United Kingdom and New York, NY, USA: Cambridge University Press.

Alova, G., P. A. Trotter, and A. Money. 2021. A machine-learning approach to predicting Africa's electricity mix based on planned power plants and their chances of success. *Nature Energy* 6(2):158-166. https://doi.org/10.1038/s41560-020-00755-9.

Alvarez, R. A., D. Zavala-Araiza, D. R. Lyon, D. T. Allen, Z. R. Barkley, A. R. Brandt, K. J. Davis, S. C. Herndon, D. J. Jacob, A. Karion, E. A. Kort, B. K. Lamb, T. Lauvaux, J. D. Maasakkers, A. J. Marchese, M. Omara, S. W. Pacala, J. Peischl, A. L. Robinson, P. B. Shepson, C. Sweeney, A. Townsend-Small, S. C. Wofsy, and S. P. Hamburg. 2018. Assessment of methane emissions from the U.S. oil and gas supply chain. *Science* 361(6398):186-188. https://doi.org/10.1126/science.aar7204.

Amini, S., T. Kuwayama, L. Gong, M. Falk, Y. Chen, Q. Mitloehner, S. Weller, F. M. Mitloehner, D. Patteson, S. A. Conley, E. Scheehle, and M. FitzGibbon. 2022. Evaluating California dairy methane emission factors using short-term ground-level and airborne measurements. *Atmospheric Environment: X* 14:100171. https://doi.org/10.1016/j.aeaoa.2022.100171.

Ammoura, L., I. Xueref-Remy, F. Vogel, V. Gros, A. Baudic, B. Bonsang, M. Delmotte, Y. Té, and F. Chevallier. 2016. Exploiting stagnant conditions to derive robust emission ratio estimates for CO_2, CO and volatile organic compounds in Paris. *Atmospheric Chemistry and Physics* 16(24):15653-15664. https://doi.org/10.5194/acp-16-15653-2016.

Andres, R. J., G. Marland, I. Fung, and E. Matthews. 1996. A 1° × 1° distribution of carbon dioxide emissions from fossil fuel consumption and cement manufacture, 1950–1990. *Global Biogeochemical Cycles* 10(3):419-429. https://doi.org/10.1029/96GB01523.

Andres, R. J., T. A. Boden, F. M. Bréon, P. Ciais, S. Davis, D. Erickson, J. S. Gregg, A. Jacobson, G. Marland, J. Miller, T. Oda, J. G. J. Olivier, M. R. Raupach, P. Rayner, and K. Treanton. 2012. A synthesis of carbon dioxide emissions from fossil-fuel combustion. *Biogeosciences* 9(5):1845-1871. https://doi.org/10.5194/bg-9-1845-2012.

Andres, R. J., T. A. Boden, and D. Higdon. 2014. A new evaluation of the uncertainty associated with CDIAC estimates of fossil fuel carbon dioxide emission. *Tellus B: Chemical and Physical Meteorology* 66(1):23616. https://doi.org/10.3402/tellusb.v66.23616.

Andres, R. J., T. A. Boden, and D. M. Higdon. 2016. Gridded uncertainty in fossil fuel carbon dioxide emission maps, a CDIAC example. *Atmospheric Chemistry and Physics* 16(23):14979-14995. https://doi.org/10.5194/acp-16-14979-2016.

Andrew, R. M. 2020. A comparison of estimates of global carbon dioxide emissions from fossil carbon sources. *Earth System Science Data* 12(2):1437-1465. https://doi.org/10.5194/essd-12-1437-2020.

Andrews, A. 2022. Developing greenhouse gas emissions inventories. Presented at Development of a Framework for Evaluating Greenhouse Gas Emissions Information for Decision Making: A Workshop, June 27, 2022.

Arnautu, D., and C. Dagenais. 2021. Use and effectiveness of policy briefs as a knowledge transfer tool: A scoping review. *Humanities and Social Sciences Communications* 8(1):211. https://doi.org/10.1057/s41599-021-00885-9.

Arnold, T., A. J. Manning, J. Kim, S. Li, H. Webster, D. Thomson, J. Mühle, R. F. Weiss, S. Park, and S. O'Doherty. 2018. Inverse modelling of CF_4 and NF_3 emissions in East Asia. *Atmospheric Chemistry and Physics* 18(18):13305-13320. https://doi.org/10.5194/acp-18-13305-2018.

Asefi-Najafabady, S., P. J. Rayner, K. R. Gurney, A. McRobert, Y. Song, K. Coltin, J. Huang, C. Elvidge, and K. Baugh. 2014. A multiyear, global gridded fossil fuel CO_2 emission data product: Evaluation and analysis of results. *Journal of Geophysical Research: Atmospheres* 119(17):10213-10231. https://doi.org/10.1002/2013JD021296.

Balkovič, J., M. Madaras, R. Skalský, C. Folberth, M. Smatanová, E. Schmid, M. van der Velde, F. Kraxner, and M. Obersteiner. 2020. Verifiable soil organic carbon modelling to facilitate regional reporting of cropland carbon change: A test case in the Czech Republic. *Journal of Environmental Management* 274:111206. https://doi.org/10.1016/j.jenvman.2020.111206.

Bastos, A., K. Hartung, T. B. Nützel, J. E. M. S. Nabel, R. A. Houghton, and J. Pongratz. 2021. Comparison of uncertainties in land-use change fluxes from bookkeeping model parameterisation. *Earth System Dynamics* 12(2):745-762. https://doi.org/10.5194/esd-12-745-2021.

Basu, S., J. B. Miller, and S. Lehman. 2016. Separation of biospheric and fossil fuel fluxes of CO_2 by atmospheric inversion of CO_2 and $^{14}CO_2$ measurements: Observation system simulations. *Atmospheric Chemistry and Physics* 16(9):5665-5683. https://doi.org/10.5194/acp-16-5665-2016.

Basu, S., D. F. Baker, F. Chevallier, P. K. Patra, J. Liu, and J. B. Miller. 2018. The impact of transport model differences on CO_2 surface flux estimates from OCO-2 retrievals of column average CO_2. *Atmospheric Chemistry and Physics* 18(10):7189-7215. https://doi.org/10.5194/acp-18-7189-2018.

REFERENCES

Basu, S., S. J. Lehman, J. B. Miller, A. E. Andrews, C. Sweeney, K. R. Gurney, X. Xu, J. Southon, and P. P. Tans. 2020. Estimating US fossil fuel CO_2 emissions from measurements of ^{14}C in atmospheric CO_2. *Proceedings of the National Academy of Sciences* 117(24):13300-13307. https://doi.org/10.1073/pnas.1919032117.

Battle, M., M. L. Bender, P. P. Tans, J. W. White, J. T. Ellis, T. Conway, and R. J. Francey. 2000. Global carbon sinks and their variability inferred from atmospheric O_2 and $\delta^{13}C$. *Science* 287(5462):2467-2470. https://doi.org/10.1126/science.287.5462.2467.

Belcher, O., P. Bigger, B. Neimark, and C. Kennelly. 2020. Hidden carbon costs of the "everywhere war": Logistics, geopolitical ecology, and the carbon boot-print of the US military. *Transactions of the Institute of British Geographers* 45(1):65-80. https://doi.org/10.1111/tran.12319.

Ben-Sasson, E., A. Chiesa, M. Green, E. Tromer, and M. Virza. 2015. Secure sampling of public parameters for succinct zero knowledge proofs. Presented at 2015 IEEE Symposium on Security and Privacy, May 17-21.

Bergamaschi, P., U. Karstens, A. J. Manning, M. Saunois, A. Tsuruta, A. Berchet, A. T. Vermeulen, T. Arnold, G. Janssens-Maenhout, S. Hammer, I. Levin, M. Schmidt, M. Ramonet, M. Lopez, J. Lavric, T. Aalto, H. Chen, D. G. Feist, C. Gerbig, L. Haszpra, O. Hermansen, G. Manca, J. Moncrieff, F. Meinhardt, J. Necki, M. Galkowski, S. O'Doherty, N. Paramonova, H. A. Scheeren, M. Steinbacher, and E. Dlugokencky. 2018. Inverse modelling of European CH_4 emissions during 2006–2012 using different inverse models and reassessed atmospheric observations. *Atmospheric Chemistry and Physics* 18(2):901-920. https://doi.org/10.5194/acp-18-901-2018.

Bertoldi, P., A. Kona, S. Rivas, and J. F. Dallemand. 2018. Towards a global comprehensive and transparent framework for cities and local governments enabling an effective contribution to the Paris climate agreement. *Current Opinion in Environmental Sustainability* 30:67-74. https://doi.org/10.1016/j.cosust.2018.03.009.

Boden, T. A., G. Marland, and R. J. Andres. 2011. *Global, Regional, and National Fossil-Fuel CO_2 Emissions*. Oak Ridge, TN: Carbon Dioxide Information Analysis Center, Oak Ridge National Laboratory, U.S. Department of Energy. http://doi.org/10.3334/CDIAC/00001_V2011.

Buolamwini, J., and T. Gebru. 2018. *Gender Shades: Intersectional Accuracy Disparities in Commercial Gender Classification*. Presented at Proceedings of the 1st Conference on Fairness, Accountability and Transparency, Proceedings of Machine Learning Research.

Bousquet, P., P. Peylin, P. Ciais, C. L. Quéré, P. Friedlingstein, and P. P. Tans. 2000. Regional Changes in Carbon Dioxide Fluxes of Land and Oceans Since 1980. *Science* 290(5495):1342-1346. https://doi.org/10.1126/science.290.5495.1342.

BP. 2022. *BP Statistical Review of World Energy 2022*. 71st edition. London: BP p.l.c. https://www.bp.com/content/dam/bp/business-sites/en/global/corporate/pdfs/energy-economics/statistical-review/bp-stats-review-2022-full-report.pdf.

Brantley, H. L., E. D. Thoma, W. C. Squier, B. B. Guven, and D. Lyon. 2014. Assessment of methane emissions from oil and gas production pads using mobile measurements. *Environmental Science & Technology* 48(24):14508-14515. https://doi.org/10.1021/es503070q.

Brasseur, G. P., and M. Gupta. 2010. Impact of aviation on climate: Research priorities. *Bulletin of the American Meteorological Society* 91(4):461-464. https://doi.org/10.1175/2009BAMS2850.1.

Brazzola, N., A. Patt, and J. Wohland. 2022. Definitions and implications of climate-neutral aviation. *Nature Climate Change* 12:761-767. https://doi.org/10.1038/s41558-022-01404-7.

Breidenich, C. 2011. *Improving Reporting of National Communications and GHG Inventories by Non-Annex I Parties under the Climate Convention*. New York: Natural Resources Defense Council. https://www.nrdc.org/sites/default/files/trackingcarbon-wp.pdf.

Brewer, P. J., R. J. C. Brown, O. A. Tarasova, B. Hall, G. C. Rhoderick, and R. I. Wielgosz. 2018. SI traceability and scales for underpinning atmospheric monitoring of greenhouse gases. *Metrologia* 55(5):S174-S181. https://doi.org/10.1088/1681-7575/aad830.

Bun, R., Z. Nahorski, J. Horabik-Pyzel, O. Danylo, L. See, N. Charkovska, P. Topylko, M. Halushchak, M. Lesiv, M. Valakh, and V. Kinakh. 2019. Development of a high-resolution spatial inventory of greenhouse gas emissions for Poland from stationary and mobile sources. *Mitigation and Adaptation Strategies for Global Change* 24(6):853-880. https://doi.org/10.1007/s11027-018-9791-2.

Bun, R. 2022. GHG emissions high-resolution spatial estimates: Uncertainty quantification. Presented at Development of a Framework for Evaluating Greenhouse Gas Emissions Information for Decision Making: A Workshop, Washington, DC, June 27–28, 2022.

Bun, R., M. Gusti, L. Kujii, O. Tokar, Y. Tsybrivskyy, and A. Bun. 2007. Spatial GHG inventory: Analysis of uncertainty sources. A case study for Ukraine. In *Accounting for Climate Change: Uncertainty in Greenhouse Gas Inventories – Verification, Compliance, and Trading*. D. Lieberman, M. Jonas, Z. Nahorski, and S. Nilsson, eds. Dordrecht: Springer Netherlands.

Burns, K., L. A. Hendricks, K. Saenko, T. Darrell, and A. Rohrbach. 2019. Women also snowboard: Overcoming bias in captioning models. arXiv 1803.09797.

Byrne, B., D. F. Baker, S. Basu, M. Bertolacci, K. W. Bowman, D. Carroll, A. Chatterjee, F. Chevallier, P. Ciais, N. Cressie, D. Crisp, S. Crowell, F. Deng, Z. Deng, N. M. Deutscher, M. Dubey, S. Feng, O. García, D. W. T. Griffith, B. Herkommer, L. Hu, A. R. Jacobson, R. Janardanan, S. Jeong, M. S. Johnson, D. B. A. Jones, R. Kivi, J. Liu, Z. Liu, S. Maksyutov, J. B. Miller, S. M. Miller, I. Morino, J. Notholt, T. Oda, C. W. O'Dell, Y. S. Oh, H. Ohyama, P. K. Patra, H. Peiro, C. Petri, S. Philip, D. F. Pollard, B. Poulter, M. Remaud, A. Schuh, M. K. Sha, K. Shiomi, K. Strong, C. Sweeney, Y. Té, H. Tian, V. A. Velazco, M. Vrekoussis, T. Warneke, J. R. Worden, D. Wunch, Y. Yao, J. Yun, A. Zammit-Mangion, and N. Zeng. 2022. National CO_2 budgets (2015–2020) inferred from atmospheric CO_2 observations in support of the Global Stocktake. *Earth System Science Data Discussions* 2022:1-59. https://doi.org/10.5194/essd-2022-213.

Cambaliza, M. O. L., P. B. Shepson, D. R. Caulton, B. Stirm, D. Samarov, K. R. Gurney, J. Turnbull, K. J. Davis, A. Possolo, A. Karion, C. Sweeney, B. Moser, A. Hendricks, T. Lauvaux, K. Mays, J. Whetstone, J. Huang, I. Razlivanov, N. L. Miles, and S. J. Richardson. 2014. Assessment of uncertainties of an aircraft-based mass balance approach for quantifying urban greenhouse gas emissions. *Atmosperic Chemistry and Physics* 14(17):9029-9050. https://doi.org/10.5194/acp-14-9029-2014.

Cambaliza, M. O. L., P. B. Shepson, J. Bogner, D. R. Caulton, B. Stirm, C. Sweeney, S. A. Montzka, K. R. Gurney, K. Spokas, O. E. Salmon, T. N. Lavoie, A. Hendricks, K. Mays, J. Turnbull, B. R. Miller, T. Lauvaux, K. Davis, A. Karion, B. Moser, C. Miller, C. Obermeyer, J. Whetstone, K. Prasad, N. Miles, and S. Richardson. 2015. Quantification and source apportionment of the methane emission flux from the city of Indianapolis. *Elementa: Science of the Anthropocene* 3. https://doi.org/10.12952/journal.elementa.000037.

Cambaliza, M. O. L., J. E. Bogner, R. B. Green, P. B. Shepson, T. A. Harvey, K. A. Spokas, B. H. Stirm, and M. Corcoran. 2017. Field measurements and modeling to resolve m^2 to km^2 CH_4 emissions for a complex urban source: An Indiana landfill study. *Elementa: Science of the Anthropocene* 5. https://doi.org/10.1525/elementa.145.

Cannon, C., S. Greene, T. K. Blank, J. Lee, and P. Natali. 2020. *The next frontier of carbon accounting: A unified approach for unlocking systemic change.* Basalt, CO: Rocky Mountain Institute. https://rmi.org/insight/the-next-frontier-of-carbon-accounting/.

Carlson, D., and T. Oda. 2018. Editorial: Data publication – ESSD goals, practices and recommendations. *Earth System Science Data* 10(4):2275-2278. https://doi.org/10.5194/essd-10-2275-2018.

Carranza, V., B. Biggs, D. Meyer, A. Townsend-Small, R. R. Thiruvenkatachari, A. Venkatram, M. L. Fischer, and F. M. Hopkins. 2022. Isotopic signatures of methane emissions from dairy farms in California's San Joaquin Valley. *Journal of Geophysical Research: Biogeosciences* 127(1):e2021JG006675. https://doi.org/10.1029/2021JG006675.

Castelvecchi, D. 2016. Can we open the black box of AI? *Nature News* 538:20-23. https://doi.org/10.1038/538020a.

CEOS (Committee on Earth Observation Satellites). 2018. A Constellation Architecture for Monitoring Carbon Dioxide and Methane from Space. https://ceos.org/observations/documents/CEOS_AC-VC_GHG_White_Paper_Publication_Draft2_20181111.pdf.

Chan, E., D. E. J. Worthy, D. Chan, M. Ishizawa, M. D. Moran, A. Delcloo, and F. Vogel. 2020. Eight-year estimates of methane emissions from oil and gas operations in western Canada are nearly twice those reported in inventories. *Environmental Science & Technology* 54(23):14899-14909. https://doi.org/10.1021/acs.est.0c04117.

Cheewaphongphan, P., S. Chatani, and N. Saigusa. 2019. Exploring gaps between bottom-up and top-down emission estimates based on uncertainties in multiple emission inventories: A case study on CH_4 emissions in China. *Sustainability* 11(7):2054. https://doi.org/10.3390/su11072054.

Chen, F., P. Deng, J. Wan, D. Zhang, A. V. Vasilakos, and X. Rong. 2015. Data mining for the Internet of Things: Literature review and challenges. *International Journal of Distributed Sensor Networks* 11(8):431047. https://doi.org/10.1155/2015/431047.

Chen, H. W., L. N. Zhang, F. Zhang, K. J. Davis, T. Lauvaux, S. Pal, B. Gaudet, and J. P. DiGangi. 2019. Evaluation of regional CO_2 mole fractions in the ECMWF CAMS real-time atmospheric analysis and NOAA CarbonTracker Near-Real-Time reanalysis with airborne observations from ACT-America field campaigns. *Journal of Geophysical Research: Atmospheres* 124(14):8119-8133. https://doi.org/10.1029/2018JD029992.

Chen, J., and M. D. Brauch. 2021. *Comparison Between the IPCC Reporting Framework and Country Practice.* New York: Columbia Center on Sustainable Investment. https://doi.org/10.7916/d8-157j-f013.

Chen, Q., M. Modi, G. McGaughey, Y. Kimura, E. McDonald-Buller, and D. T. Allen. 2022a. Simulated methane emission detection capabilities of continuous monitoring networks in an oil and gas production region. *Atmosphere* 13:510. https://doi.org/10.3390/atmos13040510.

Chen, Y., E. D. Sherwin, E. S. F. Berman, B. B. Jones, M. P. Gordon, E. B. Wetherley, E. A. Kort, and A. R. Brandt. 2022b. Quantifying regional methane emissions in the New Mexico Permian Basin with a comprehensive aerial survey. *Environmental Science & Technology* 56(7):4317-4323. https://doi.org/10.1021/acs.est.1c06458.

Chevallier, F., B. Zheng, G. Broquet, P. Ciais, Z. Liu, S. J. Davis, Z. Deng, Y. Wang, F.-M. Bréon, and C. W. O'Dell. 2020. Local anomalies in the column-averaged dry air mole fractions of carbon dioxide across the globe during the first months of the coronavirus recession. *Geophysical Research Letters* 47(22):e2020GL090244. https://doi.org/10.1029/2020GL090244.

REFERENCES

Ciais, P. 2015. Decadal changes in carbon emissions and sinks optimized using a Bayesian fusion of multiple observations. Presented at American Geophysical Union Fall Meeting, December 2015.

Ciais, P., P. P. Tans, M. Trolier, J. W. C. White, and R. J. Francey. 1995. A large northern hemisphere terrestrial CO_2 sink indicated by the $^{13}C/^{12}C$ ratio of atmospheric CO_2. *Science* 269(5227):1098-1102. https://doi.org/10.1126/science.269.5227.1098.

ClimaSouth. 2014. *An Introduction to National GHG Inventories Measurement, Reporting & Verification (MRV)*. Rabat, Morocco: ClimaSouth. https://www.climamed.eu/wp-content/uploads/files/An-Introduction-to-National-GHG-Inventories.pdf.

Climate TRACE. n.d. Climate Trace: Independent Greenhouse Gas Emissions Tracking. https://www.climatetrace.org.

Conley, S., I. Faloona, S. Mehrotra, M. Suard, D. H. Lenschow, C. Sweeney, S. Herndon, S. Schwietzke, G. Pétron, J. Pifer, E. A. Kort, and R. Schnell. 2017. Application of Gauss's theorem to quantify localized surface emissions from airborne measurements of wind and trace gases. *Atmospheric Measurement Techniques* 10(9):3345-3358. https://doi.org/10.5194/amt-10-3345-2017.

Conrad, B. M., and M. R. Johnson. 2017. Field measurements of black carbon yields from gas flaring. *Environmental Science & Technology* 51(3):1893-1900. https://doi.org/10.1021/acs.est.6b03690.

Crippa, M., D. Guizzardi, E. Solazzo, M. Muntean, E. Schaaf, F. Monforti-Ferrario, M. Banja, J. G. J. Olivier, G. Grassi, S. Rossi, and E. Vignati. 2021. *GHG Emissions of All World Countries*. 2021 Report, EUR 30831 EN. Luxembourg: Publications Office of the European Union.

Crowell, S., D. Baker, A. Schuh, S. Basu, A. R. Jacobson, F. Chevallier, J. Liu, F. Deng, L. Feng, K. McKain, A. Chatterjee, J. B. Miller, B. B. Stephens, A. Eldering, D. Crisp, D. Schimel, R. Nassar, C. W. O'Dell, T. Oda, C. Sweeney, P. I. Palmer, and D. B. A. Jones. 2019. The 2015–2016 carbon cycle as seen from OCO-2 and the global in situ network. *Atmospheric Chemistry and Physics* 19(15):9797-9831. https://doi.org/10.5194/acp-19-9797-2019.

Cui, Y. Y., J. Brioude, W. M. Angevine, J. Peischl, S. A. McKeen, S.-W. Kim, J. A. Neuman, D. K. Henze, N. Bousserez, M. L. Fischer, S. Jeong, H. A. Michelsen, R. P. Bambha, Z. Liu, G. W. Santoni, B. C. Daube, E. A. Kort, G. J. Frost, T. B. Ryerson, S. C. Wofsy, and M. Trainer. 2017. Top-down estimate of methane emissions in California using a mesoscale inverse modeling technique: The San Joaquin Valley. *Journal of Geophysical Research: Atmospheres* 122(6):3686-3699. https://doi.org/10.1002/2016JD026398.

Cui, Y. Y., A. Vijayan, M. Falk, Y.-K. Hsu, D. Yin, X. M. Chen, Z. Zhao, J. Avise, Y. Chen, K. Verhulst, R. Duren, V. Yadav, C. Miller, R. Weiss, R. Keeling, J. Kim, L. T. Iraci, T. Tanaka, M. S. Johnson, E. A. Kort, L. Bianco, M. L. Fischer, K. Stroud, J. Herner, and B. Croes. 2019. A multiplatform inversion estimation of statewide and regional methane emissions in California during 2014–2016. *Environmental Science & Technology* 53(16):9636-9645. https://doi.org/10.1021/acs.est.9b01769.

Cusworth, D. H., R. M. Duren, A. K. Thorpe, E. Tseng, D. Thompson, A. Guha, S. Newman, K. T. Foster, and C. E. Miller. 2020. Using remote sensing to detect, validate, and quantify methane emissions from California solid waste operations. *Environmental Research Letters* 15(5):054012. https://doi.org/10.1088/1748-9326/ab7b99.

Cusworth, D. H., R. M. Duren, A. K. Thorpe, M. L. Eastwood, R. O. Green, P. E. Dennison, C. Frankenberg, J. W. Heckler, G. P. Asner, and C. E. Miller. 2021a. Quantifying global power plant carbon dioxide emissions with imaging spectroscopy. *AGU Advances* 2(2):e2020AV000350. https://doi.org/10.1029/2020AV000350.

Cusworth, D. H., R. M. Duren, A. K. Thorpe, W. Olson-Duvall, J. Heckler, J. W. Chapman, M. L. Eastwood, M. C. Helmlinger, R. O. Green, G. P. Asner, P. E. Dennison, and C. E. Miller. 2021b. Intermittency of large methane emitters in the Permian Basin. *Environmental Science & Technology Letters* 8(7):567-573. https://doi.org/10.1021/acs.estlett.1c00173.

Damassa, T., and S. Elsayed. 2013. *From the GHG Measurement Frontline: A Synthesis of Non-Annex I Country National Inventory System Practices and Experiences*. Washington, DC: World Resource Institute. https://www.wri.org/research/ghg-measurement-frontline.

Davis, K. J., A. Deng, T. Lauvaux, N. L. Miles, S. J. Richardson, D. P. Sarmiento, K. R. Gurney, R. M. Hardesty, T. A. Bonin, W. A. Brewer, B. K. Lamb, P. B. Shepson, R. M. Harvey, M. O. Cambaliza, C. Sweeney, J. C. Turnbull, J. Whetstone, and A. Karion. 2017. The Indianapolis Flux Experiment (INFLUX): A test-bed for developing urban greenhouse gas emission measurements. *Elementa: Science of the Anthropocene*. https://doi.org/10.1525/elementa.188.

Davis, S. J., and K. Caldeira. 2010. Consumption-based accounting of CO_2 emissions. *Proceedings of the National Academy of Sciences* 107(12):5687-5692. https://doi.org/10.1073/pnas.0906974107.

Day, T., S. Mooldijk, S. Smit, E. Posada, F. Hans, H. Fearnehough, A. Kachi, C. Warnecke, T. Kuramochi, and N. Höhne. 2022. *Corporate Climate Responsibility Monitor 2022*. Cologne, Germany: NewClimate Institute. https://newclimate.org/resources/publications/corporate-climate-responsibility-monitor-2022.

De Filippi, P., M. Mannan, and W. Reijers. 2020. Blockchain as a confidence machine: The problem of trust & challenges of governance. *Technology in Society* 62:101284. https://doi.org/10.1016/j.techsoc.2020.101284.

Deng, A., T. Lauvaux, K. J. Davis, B. J. Gaudet, N. Miles, S. J. Richardson, K. Wu, D. P. Sarmiento, R. M. Hardesty, T. A. Bonin, W. A. Brewer, and K. R. Gurney. 2017. Toward reduced transport errors in a high resolution urban CO_2 inversion system. *Elementa: Science of the Anthropocene* 5. https://doi.org/10.1525/elementa.133.

Deng, Z., P. Ciais, Z. A. Tzompa-Sosa, M. Saunois, C. Qiu, C. Tan, T. Sun, P. Ke, Y. Cui, K. Tanaka, X. Lin, R. L. Thompson, H. Tian, Y. Yao, Y. Huang, R. Lauerwald, A. K. Jain, X. Xu, A. Bastos, S. Sitch, P. I. Palmer, T. Lauvaux, A. d'Aspremont, C. Giron, A. Benoit, B. Poulter, J. Chang, A. M. R. Petrescu, S. J. Davis, Z. Liu, G. Grassi, C. Albergel, F. N. Tubiello, L. Perugini, W. Peters, and F. Chevallier. 2022. Comparing national greenhouse gas budgets reported in UNFCCC inventories against atmospheric inversions. *Earth System Science Data* 14(4):1639-1675. https://doi.org/10.5194/essd-14-1639-2022.

Denning, A. S., I. Y. Fung, and D. Randall. 1995. Latitudinal gradient of atmospheric CO_2 due to seasonal exchange with land biota. *Nature* 376(6537):240-243. https://doi.org/10.1038/376240a0.

Deroubaix, A., G. Brasseur, B. Gaubert, I. Labuhn, L. Menut, G. Siour, and P. Tuccella. 2021. Response of surface ozone concentration to emission reduction and meteorology during the COVID-19 lockdown in Europe. *Meteorological Applications* 28(3):e1990. https://doi.org/10.1002/met.1990.

Dietz, T., E. Ostrom, and P. C. Stern. 2003. The struggle to govern the commons. *Science* 302(5652):1907-1912. https://doi.org/10.1126/science.1091015.

Dou, X., Y. Wang, P. Ciais, F. Chevallier, S. J. Davis, M. Crippa, G. Janssens-Maenhout, D. Guizzardi, E. Solazzo, F. Yan, D. Huo, B. Zheng, B. Zhu, D. Cui, P. Ke, T. Sun, H. Wang, Q. Zhang, P. Gentine, Z. Deng, and Z. Liu. 2022. Near-real-time global gridded daily CO_2 emissions. *The Innovation* 3(1):100182. https://doi.org/10.1016/j.xinn.2021.100182.

Dreyfus, G. B., Y. Xu, D. T. Shindell, D. Zaelke, and V. Ramanathan. 2022. Mitigating climate disruption in time: A self-consistent approach for avoiding both near-term and long-term global warming. *Proceedings of the National Academy of Sciences* 119(22):e2123536119. https://doi.org/10.1073/pnas.2123536119.

Duren, R. M. 2022. Emerging approaches and integration of multiple data sources. Presented at Greenhouse Gas Emissions Monitoring, Inventories, and Data Integration: Understanding the Landscape. National Academies of Sciences, Engineering, and Medicine, Board on Atmospheric Sciences and Climate, Washington, DC, June 2.

Duren, R. M., A. K. Thorpe, K. T. Foster, T. Rafiq, F. M. Hopkins, V. Yadav, B. D. Bue, D. R. Thompson, S. Conley, N. K. Colombi, C. Frankenberg, I. B. McCubbin, M. L. Eastwood, M. Falk, J. D. Herner, B. E. Croes, R. O. Green, and C. E. Miller. 2019. California's methane super-emitters. *Nature* 575(7781):180-184. https://doi.org/10.1038/s41586-019-1720-3.

Dvorak, M. T., K. C. Armour, D. M. W. Frierson, C. Proistosescu, M. B. Baker, and C. J. Smith. 2022. Estimating the timing of geophysical commitment to 1.5 and 2.0 °C of global warming. *Nature Climate Change* 12(6):547-552. https://doi.org/10.1038/s41558-022-01372-y.

Eggleton, F., and K. Winfield. 2020. Open data challenges in climate science. *Data Science Journal* 19(1):52. https://doi.org/10.5334/dsj-2020-052.

EIA (Energy Information Administration). 2021. *Annual Energy Outlook 2021 with Projections to 2050*. Washington, DC: U.S. Department of Energy. https://www.eia.gov/outlooks/archive/aeo21/pdf/AEO_Narrative_2021.pdf.

Eilperin, J., and C. Mooney. 2021. Countries' climate pledges built on flawed data, Post investigation finds. *The Washington Post*, November 8. https://www.washingtonpost.com/climate-environment/interactive/2021/greenhouse-gas-emissions-pledges-data.

Enting, I. G. 2001. Atmospheric Inversions: Carbon Isotopes. Presented at WOCE/JGOFS Ocean Transport Workshop, Southampton, UK, June 25-29. https://www.nodc.noaa.gov/archive/arc0013/0001873/1.1/data/1-data/publications/WOCE/transport_wkshop.pdf#page=39.

EPA (U.S. Environmental Protection Agency). 2009. Mandatory Reporting of Greenhouse Gases. 40 CFR Parts 86, 87, 89, 90, 94, 98, 1033, 1039, 1042, 1045, 1048, 1051, 1054, 1065 [EPA–HQ–OAR–2008–0508; FRL–8963–5] RIN 2060–A079. https://www.epa.gov/sites/default/files/2015-06/documents/ghg-mrr-finalpreamble.pdf.

EPA. 2019. Inventory of U.S. Greenhouse Gas Emissions and Sinks: 1990-2017. EPA 430-R-19-001. https://www.epa.gov/sites/default/files/2019-04/documents/us-ghg-inventory-2019-main-text.pdf.

EPA. 2021. Standards of Performance for New, Reconstructed, and Modified Sources and Emissions Guidelines for Existing Sources: Oil and Natural Gas Sector Climate Review. 40 CFR Part 60 [EPA-HQ-OAR-2021-0317; FRL-8510-02-OAR] RIN 2060-AV16. https://www.epa.gov/system/files/documents/2021-11/san-8510-ong-climate-review-proposal-frn-2021-11_1.pdfw.

EPA. 2022. Inventory of U.S. Greenhouse Gas Emissions and Sinks: 1990-2020. EPA 430-R-22-003. https://www.epa.gov/ghgemissions/draft-inventory-us-greenhouse-gas-emissions-and-sinks-1990-2020.

European Commission. 2022. EU Taxonomy for Sustainable Activities. https://ec.europa.eu/info/business-economy-euro/banking-and-finance/sustainable-finance/eu-taxonomy-sustainable-activities_en.

Executive Office of the President. 2021. *The Long-Term Strategy of the United States: Pathways to Net-Zero Greenhouse Gas Emissions by 2050*. Washington, DC: U.S. Department of State and the U.S. Executive Office of the President. https://www.whitehouse.gov/wp-content/uploads/2021/10/US-Long-Term-Strategy.pdf.

Falduto, C., and S. Wartmann. 2021. Towards common GHG inventory reporting tables for Biennial Transparency Reports: Experiences with tools for generating and using reporting tables under the UNFCCC. *OECD/IEA Climate Change Expert Group Papers* 2021(01). https://doi.org/10.1787/38f54dbf-en.

Fernández-Martínez, M., J. Sardans, F. Chevallier, P. Ciais, M. Obersteiner, S. Vicca, J. G. Canadell, A. Bastos, P. Friedlingstein, S. Sitch, S. L. Piao, I. A. Janssens, and J. Peñuelas. 2019. Global trends in carbon sinks and their relationships with CO_2 and temperature. *Nature Climate Change* 9(1):73-79. https://doi.org/10.1038/s41558-018-0367-7.

Fischer, M. L., N. Parazoo, K. Brophy, X. Cui, S. Jeong, J. Liu, R. Keeling, T. E. Taylor, K. Gurney, T. Oda, and H. Graven. 2017. Simulating estimation of California fossil fuel and biosphere carbon dioxide exchanges combining in situ tower and satellite column observations. *Journal of Geophysical Research: Atmospheres* 122(6):3653-3671. https://doi.org/10.1002/2016JD025617.

Fischer, M. L., E. D. Lebel, and R. B. Jackson. 2020. *Quantifying Methane from California's Plugged and Abandoned Oil and Gas Wells*. Sacramento, CA: California Energy Commission. https://www.energy.ca.gov/sites/default/files/2021-05/CEC-500-2020-052.pdf.

Fitzmaurice, H. L., A. J. Turner, J. Kim, K. Chan, E. R. Delaria, C. Newman, P. Wooldridge, and R. C. Cohen. 2022. Assessing vehicle fuel efficiency using a dense network of CO_2 observations. *Atmospheric Chemistry and Physics* 22(6):3891-3900. https://doi.org/10.5194/acp-22-3891-2022.

Flerlage, H., G. J. M. Velders, and J. de Boer. 2021. A review of bottom-up and top-down emission estimates of hydrofluorocarbons (HFCs) in different parts of the world. *Chemosphere* 283:131208. https://doi.org/10.1016/j.chemosphere.2021.131208.

Fong, W. K., M. Sotos, M. Doust, S. Schultz, A. Marques, and C. Deng-Beck. 2014. Global Protocol for Community-Scale Greenhouse Gas Emission Inventories. Washington, DC: World Resources Institute. https://ghgprotocol.org/greenhouse-gas-protocol-accounting-reporting-standard-cities.

Forster, P. M., H. I. Forster, M. J. Evans, M. J. Gidden, C. D. Jones, C. A. Keller, R. D. Lamboll, C. L. Quéré, J. Rogelj, D. Rosen, C.-F. Schleussner, T. B. Richardson, C. J. Smith, and S. T. Turnock. 2020. Current and future global climate impacts resulting from COVID-19. *Nature Climate Change* 10(10):913-919. https://doi.org/10.1038/s41558-020-0883-0.

Fournier, J., V. Gray, and J. Grippe. 2022. JLINX White Paper.

Frankenberg, C., S. S. Kulawik, S. C. Wofsy, F. Chevallier, B. Daube, E. A. Kort, C. O'Dell, E. T. Olsen, and G. Osterman. 2016. Using airborne HIAPER Pole-to-Pole Observations (HIPPO) to evaluate model and remote sensing estimates of atmospheric carbon dioxide. *Atmospheric Chemistry and Physics* 16(12):7867-7878. https://doi.org/10.5194/acp-16-7867-2016.

Frey, M., M. K. Sha, F. Hase, M. Kiel, T. Blumenstock, R. Harig, G. Surawicz, N. M. Deutscher, K. Shiomi, J. E. Franklin, H. Bösch, J. Chen, M. Grutter, H. Ohyama, Y. Sun, A. Butz, G. Mengistu Tsidu, D. Ene, D. Wunch, Z. Cao, O. Garcia, M. Ramonet, F. Vogel, and J. Orphal. 2019. Building the COllaborative Carbon Column Observing Network (COCCON): Long-term stability and ensemble performance of the EM27/SUN Fourier transform spectrometer. *Atmospheric Measurement Techniques* 12(3):1513-1530. https://doi.org/10.5194/amt-12-1513-2019.

Frey, M. M., F. Hase, T. Blumenstock, D. Dubravica, J. Groß, F. Göttsche, M. Handjaba, P. Amadhila, R. Mushi, I. Morino, K. Shiomi, M. K. Sha, M. de Mazière, and D. F. Pollard. 2021. Long-term column-averaged greenhouse gas observations using a COCCON spectrometer at the high-surface-albedo site in Gobabeb, Namibia. *Atmospheric Measurement Techniques* 14(9):5887-5911. https://doi.org/10.5194/amt-14-5887-2021.

Friedlingstein, P., M. W. Jones, M. O'Sullivan, R. M. Andrew, J. Hauck, G. P. Peters, W. Peters, J. Pongratz, S. Sitch, C. Le Quéré, D. C. E. Bakker, J. G. Canadell, P. Ciais, R. B. Jackson, P. Anthoni, L. Barbero, A. Bastos, V. Bastrikov, M. Becker, L. Bopp, E. Buitenhuis, N. Chandra, F. Chevallier, L. P. Chini, K. I. Currie, R. A. Feely, M. Gehlen, D. Gilfillan, T. Gkritzalis, D. S. Goll, N. Gruber, S. Gutekunst, I. Harris, V. Haverd, R. A. Houghton, G. Hurtt, T. Ilyina, A. K. Jain, E. Joetzjer, J. O. Kaplan, E. Kato, K. Klein Goldewijk, J. I. Korsbakken, P. Landschützer, S. K. Lauvset, N. Lefèvre, A. Lenton, S. Lienert, D. Lombardozzi, G. Marland, P. C. McGuire, J. R. Melton, N. Metzl, D. R. Munro, J. E. M. S. Nabel, S. I. Nakaoka, C. Neill, A. M. Omar, T. Ono, A. Peregon, D. Pierrot, B. Poulter, G. Rehder, L. Resplandy, E. Robertson, C. Rödenbeck, R. Séférian, J. Schwinger, N. Smith, P. P. Tans, H. Tian, B. Tilbrook, F. N. Tubiello, G. R. van der Werf, A. J. Wiltshire, and S. Zaehle. 2019. Global carbon budget 2019. *Earth System Science Data* 11(4):1783-1838. https://doi.org/10.5194/essd-11-1783-2019.

Friedlingstein, P., M. O'Sullivan, M. W. Jones, R. M. Andrew, J. Hauck, A. Olsen, G. P. Peters, W. Peters, J. Pongratz, S. Sitch, C. Le Quéré, J. G. Canadell, P. Ciais, R. B. Jackson, S. Alin, L. E. O. C. Aragão, A. Arneth, V. Arora, N. R. Bates, M. Becker, A. Benoit-Cattin, H. C. Bittig, L. Bopp, S. Bultan, N. Chandra, F. Chevallier, L. P. Chini, W. Evans, L. Florentie, P. M. Forster, T. Gasser, M. Gehlen, D. Gilfillan, T. Gkritzalis, L. Gregor, N. Gruber, I. Harris, K. Hartung, V. Haverd, R. A. Houghton, T. Ilyina, A. K. Jain, E. Joetzjer, K. Kadono, E. Kato, V. Kitidis, J. I. Korsbakken, P. Landschützer, N. Lefèvre, A. Lenton, S. Lienert, Z. Liu, D. Lombardozzi, G. Marland, N. Metzl, D. R. Munro, J. E. M. S. Nabel, S. I. Nakaoka, Y. Niwa, K. O'Brien, T. Ono, P. I. Palmer, D. Pierrot, B. Poulter, L. Resplandy, E. Robertson, C. Rödenbeck, J. Schwinger, R. Séférian, I. Skjelvan, A. J. P. Smith, A. J. Sutton, T. Tanhua, P. P. Tans, H. Tian, B. Tilbrook, G. van der Werf, N. Vuichard, A. P. Walker, R. Wanninkhof, A. J. Watson, D. Willis, A. J. Wiltshire, W. Yuan, X. Yue, and S. Zaehle. 2020. Global carbon budget 2020. *Earth System Science Data* 12(4):3269-3340. https://doi.org/10.5194/essd-12-3269-2020.

Friedlingstein, P., M. W. Jones, M. O'Sullivan, R. M. Andrew, D. C. E. Bakker, J. Hauck, C. Le Quéré, G. P. Peters, W. Peters, J. Pongratz, S. Sitch, J. G. Canadell, P. Ciais, R. B. Jackson, S. R. Alin, P. Anthoni, N. R. Bates, M. Becker, N. Bellouin, L. Bopp, T. T. T. Chau, F. Chevallier, L. P. Chini, M. Cronin, K. I. Currie, B. Decharme, L. M. Djeutchouang, X. Dou, W. Evans, R. A. Feely, L. Feng, T. Gasser, D. Gilfillan, T. Gkritzalis, G. Grassi, L. Gregor, N. Gruber, Ö. Gürses, I. Harris, R. A. Houghton, G. C. Hurtt, Y. Iida, T. Ilyina, I. T. Luijkx, A. Jain, S. D. Jones, E. Kato, D. Kennedy, K. Klein Goldewijk, J. Knauer, J. I. Korsbakken, A. Körtzinger, P. Landschützer, S. K. Lauvset, N. Lefèvre, S. Lienert, J. Liu, G. Marland, P. C. McGuire, J. R. Melton, D. R. Munro, J. E. M. S. Nabel, S. I. Nakaoka, Y. Niwa, T. Ono, D. Pierrot, B. Poulter, G. Rehder, L. Resplandy, E. Robertson, C. Rödenbeck, T. M. Rosan, J. Schwinger, C. Schwingshackl, R. Séférian, A. J. Sutton, C. Sweeney, T. Tanhua, P. P. Tans, H. Tian, B. Tilbrook, F. Tubiello, G. R. van der Werf, N. Vuichard, C. Wada, R. Wanninkhof, A. J. Watson, D. Willis, A. J. Wiltshire, W. Yuan, C. Yue, X. Yue, S. Zaehle, and J. Zeng. 2022. Global carbon budget 2021. *Earth System Science Data* 14(4):1917-2005. https://doi.org/10.5194/essd-14-1917-2022.

Ganesan, A. L., A. J. Manning, A. Grant, D. Young, D. E. Oram, W. T. Sturges, J. B. Moncrieff, and S. O'Doherty. 2015. Quantifying methane and nitrous oxide emissions from the UK and Ireland using a national-scale monitoring network. *Atmospheric Chemistry and Physics* 15(11):6393-6406. https://doi.org/10.5194/acp-15-6393-2015.

Gasser, T., P. Ciais, O. Boucher, Y. Quilcaille, M. Tortora, L. Bopp, and D. Hauglustaine. 2017. The compact Earth system model OSCAR v2.2: Description and first results. *Geoscientific Model Development* 10(1):271-319. https://doi.org/10.5194/gmd-10-271-2017.

Gasser, T., L. Crepin, Y. Quilcaille, R. A. Houghton, P. Ciais, and M. Obersteiner. 2020. Historical CO_2 emissions from land use and land cover change and their uncertainty. *Biogeosciences* 17(15):4075-4101. https://doi.org/10.5194/bg-17-4075-2020.

Gately, C. K., and L. R. Hutyra. 2017. Large uncertainties in urban-scale carbon emissions. *Journal of Geophysical Research: Atmospheres* 122(20):11242-11260. https://doi.org/10.1002/2017JD027359.

Gately, C. K., L. R. Hutyra, S. Peterson, and I. Sue Wing. 2017. Urban emissions hotspots: Quantifying vehicle congestion and air pollution using mobile phone GPS data. *Environmental Pollution* 229:496-504. https://doi.org/10.1016/j.envpol.2017.05.091.

Gatti, L. V., L. S. Basso, J. B. Miller, M. Gloor, L. Gatti Domingues, H. L. G. Cassol, G. Tejada, L. E. O. C. Aragão, C. Nobre, W. Peters, L. Marani, E. Arai, A. H. Sanches, S. M. Corrêa, L. Anderson, C. Von Randow, C. S. C. Correia, S. P. Crispim, and R. A. L. Neves. 2021. Amazonia as a carbon source linked to deforestation and climate change. *Nature* 595(7867):388-393. https://doi.org/10.1038/s41586-021-03629-6.

GCoM (Global Convenant of Mayors). 2018. *Implementing Climate Ambition: Global Covenant of Mayors 2018 Global Aggregation Report*. https://www.globalcovenantofmayors.org/press/implementing-climate-ambition-global-covenant-of-mayors-2018-global-aggregation-report/. https://www.globalcovenantofmayors.org/press/implementing-climate-ambition-global-covenant-of-mayors-2018-global-aggregation-report/.

Gensheimer, J., J. Chen, A. J. Turner, A. Shekhar, A. Wenzel, and F. N. Keutsch. 2021. What are the different measures of mobility telling us about surface transportation CO_2 emissions during the COVID-19 pandemic? *Journal of Geophysical Research: Atmospheres* 126(11):e2021JD034664. https://doi.org/10.1029/2021JD034664.

Gilfillan, D., and G. Marland. 2021. CDIAC-FF: global and national CO_2 emissions from fossil fuel combustion and cement manufacture: 1751–2017. *Earth System Science Data* 13(4):1667-1680. https://doi.org/10.5194/essd-13-1667-2021.

Gioli, B., G. Gualtieri, C. Busillo, F. Calastrini, A. Zaldei, and P. Toscano. 2015. Improving high resolution emission inventories with local proxies and urban eddy covariance flux measurements. *Atmospheric Environment* 115:246-256. https://doi.org/10.1016/j.atmosenv.2015.05.068.

Gisi, M., F. Hase, S. Dohe, T. Blumenstock, A. Simon, and A. Keens. 2012. XCO_2-measurements with a tabletop FTS using solar absorption spectroscopy. *Atmospheric Measurement Techniques* 5(11):2969-2980. https://doi.org/10.5194/amt-5-2969-2012.

Giuliani, G., P. Lacroix, Y. Guigoz, R. Roncella, L. Bigagli, M. Santoro, P. Mazzetti, S. Nativi, N. Ray, and A. Lehmann. 2017. Bringing GEOSS services into practice: A capacity building resource on spatial data infrastructures (SDI). *Transactions in GIS* 21(4):811-824. https://doi.org/10.1111/tgis.12209.

Giuliani, G., J. Masó, P. Mazzetti, S. Nativi, and A. Zabala. 2019. Paving the way to increased interoperability of Earth observations data cubes. *Data* 4(3):113. https://doi.org/10.3390/data4030113.

Gommet, C., R. Lauerwald, P. Ciais, B. Guenet, H. Zhang, and P. Regnier. 2022. Spatiotemporal patterns and drivers of terrestrial dissolved organic carbon (DOC) leaching into the European river network. *Earth System Dynamics* 13(1):393-418. https://doi.org/10.5194/esd-13-393-2022.

Goodwin, J., and K. Kizzier. 2018. *Elaborating the Paris Agreement: National Greenhouse Gas Inventories*. Arlington, VA: Center for Climate and Energy Solutions. https://www.c2es.org/wp-content/uploads/2018/08/national-greenhouse-gas-inventories.pdf.

Gordon, D. 2022. Development of a Framework for Evaluating Greenhouse Gas Emissions Information for Decision Making. Presented at Workshop on Development of a Framework for Evaluating Global Greenhouse Gas Emissions Information for Decision Making. National Academies of Sciences, Engineering, and Medicine, Washington, DC, June 27.

Grange, S. K., J. D. Lee, W. S. Drysdale, A. C. Lewis, C. Hueglin, L. Emmenegger, and D. C. Carslaw. 2021. COVID-19 lockdowns highlight a risk of increasing ozone pollution in European urban areas. *Atmospheric Chemistry and Physics* 21(5):4169-4185. https://doi.org/10.5194/acp-21-4169-2021.

Grassi, G., E. Stehfest, J. Rogelj, D. van Vuuren, A. Cescatti, J. House, G.-J. Nabuurs, S. Rossi, R. Alkama, R. A. Viñas, K. Calvin, G. Ceccherini, S. Federici, S. Fujimori, M. Gusti, T. Hasegawa, P. Havlik, F. Humpenöder, A. Korosuo, L. Perugini, F. N. Tubiello, and A. Popp. 2021. Critical adjustment of land mitigation pathways for assessing countries' climate progress. *Nature Climate Change* 11(5):425-434. https://doi.org/10.1038/s41558-021-01033-6.

Graven, H., M. L. Fischer, T. Lueker, S. Jeong, T. P. Guilderson, R. F. Keeling, R. Bambha, K. Brophy, W. Callahan, X. Cui, C. Frankenberg, K. R. Gurney, B. W. LaFranchi, S. J. Lehman, H. Michelsen, J. B. Miller, S. Newman, W. Paplawsky, N. C. Parazoo, C. Sloop, and S. J. Walker. 2018. Assessing fossil fuel CO_2 emissions in California using atmospheric observations and models. *Environmental Research Letters* 13(6):065007. https://doi.org/10.1088/1748-9326/aabd43.

Greally, B. R., A. J. Manning, S. Reimann, A. McCulloch, J. Huang, B. L. Dunse, P. G. Simmonds, R. G. Prinn, P. J. Fraser, D. M. Cunnold, S. O'Doherty, L. W. Porter, K. Stemmler, M. K. Vollmer, C. R. Lunder, N. Schmidbauer, O. Hermansen, J. Arduini, P. K. Salameh, P. B. Krummel, R. H. J. Wang, D. Folini, R. F. Weiss, M. Maione, G. Nickless, F. Stordal, and R. G. Derwent. 2007. Observations of 1,1-difluoroethane (HFC-152a) at AGAGE and SOGE monitoring stations in 1994–2004 and derived global and regional emission estimates. *Journal of Geophysical Research: Atmospheres* 112(D6). https://doi.org/10.1029/2006JD007527.

Griffith, D. W. T., D. Pöhler, S. Schmitt, S. Hammer, S. N. Vardag, and U. Platt. 2018. Long open-path measurements of greenhouse gases in air using near-infrared Fourier transform spectroscopy. *Atmospheric Measurement Techniques* 11(3):1549-1563. https://doi.org/10.5194/amt-11-1549-2018.

Guan, D., Z. Liu, Y. Geng, S. Lindner, and K. Hubacek. 2012. The gigatonne gap in China's carbon dioxide inventories. *Nature Climate Change* 2(9):672-675. https://doi.org/10.1038/nclimate1560.

Gupta, A., and H. van Asselt. 2019. Transparency in multilateral climate politics: Furthering (or distracting from) accountability? *Regulation & Governance* 13(1):18-34. https://doi.org/10.1111/rego.12159.

Gurney, K., and P. Shepson. 2021. The power and promise of improved climate data infrastructure. *Proceedings of the National Academy of Sciences* 118(35):e2114115118. https://doi.org/10.1073/pnas.2114115118.

Gurney, K. R., R. M. Law, A. S. Denning, P. J. Rayner, D. Baker, P. Bousquet, L. Bruhwiler, Y.-H. Chen, P. Ciais, S. Fan, I. Y. Fung, M. Gloor, M. Heimann, K. Higuchi, J. John, T. Maki, S. Maksyutov, K. Masarie, P. Peylin, M. Prather, B. C. Pak, J. Randerson, J. Sarmiento, S. Taguchi, T. Takahashi, and C.-W. Yuen. 2002. Towards robust regional estimates of CO_2 sources and sinks using atmospheric transport models. *Nature* 415(6872):626-630. https://doi.org/10.1038/415626a.

Gurney, K. R., D. L. Mendoza, Y. Zhou, M. L. Fischer, C. C. Miller, S. Geethakumar, and S. de la Rue du Can. 2009. High resolution fossil fuel combustion CO_2 emission fluxes for the United States. *Environmental Science & Technology* 43(14):5535-5541. https://doi.org/10.1021/es900806c.

Gurney, K. R., I. Razlivanov, Y. Song, Y. Zhou, B. Benes, and M. Abdul-Massih. 2012. Quantification of fossil fuel CO_2 emissions on the building/street scale for a large U.S. city. *Environmental Science & Technology* 46(21):12194-12202. https://doi.org/10.1021/es3011282.

Gurney, K. R., P. Romero-Lankao, K. C. Seto, L. R. Hutyra, R. Duren, C. Kennedy, N. B. Grimm, J. R. Ehleringer, P. Marcotullio, S. Hughes, S. Pincetl, M. V. Chester, D. M. Runfola, J. J. Feddema, and J. Sperling. 2015. Climate change: Track urban emissions on a human scale. *Nature* 525(7568):179-181. https://doi.org/10.1038/525179a.

Gurney, K. R., J. Huang, and K. Coltin. 2016. Bias present in US federal agency power plant CO_2 emissions data and implications for the US clean power plan. *Environmental Research Letters* 11(6):064005. https://doi.org/10.1088/1748-9326/11/6/064005.

Gurney, K. R., R. Patarasuk, J. Liang, Y. Song, D. O'Keeffe, P. Rao, J. R. Whetstone, R. M. Duren, A. Eldering, and C. Miller. 2019. The Hestia fossil fuel CO_2 emissions data product for the Los Angeles megacity (Hestia-LA). *Earth System Science Data* 11(3):1309-1335. https://doi.org/10.5194/essd-11-1309-2019.

Gurney, K. R., Y. Song, J. Liang, and G. Roest. 2020a. Toward accurate, policy-relevant fossil fuel CO_2 emission landscapes. *Environmental Science & Technology* 54(16):9896-9907. https://doi.org/10.1021/acs.est.0c01175.

Gurney, K. R., J. Liang, R. Patarasuk, Y. Song, J. Huang, and G. Roest. 2020b. The Vulcan version 3.0 high-resolution fossil fuel CO_2 emissions for the United States. *Journal of Geophysical Research: Atmospheres* 125(19):e2020JD032974. https://doi.org/10.1029/2020JD032974.

Gurney, K. R., J. Liang, G. Roest, Y. Song, K. Mueller, and T. Lauvaux. 2021. Under-reporting of greenhouse gas emissions in U.S. cities. *Nature Communications* 12(1):553. https://doi.org/10.1038/s41467-020-20871-0.

Gurney, K. R., Ş. Kılkış, K. C. Seto, S. Lwasa, D. Moran, K. Riahi, M. Keller, P. Rayner, and M. Luqman. 2022. Greenhouse gas emissions from global cities under SSP/RCP scenarios, 1990 to 2100. *Global Environmental Change* 73:102478. https://doi.org/10.1016/j.gloenvcha.2022.102478.

Gütschow, J., L. Jeffery, R. Gieseke, and A. Günther. 2019. *The PRIMAP-hist National Historical Emissions Time Series (1850–2017)*. V. 2.1. Potsdam: GFZ Data Services. https://doi.org/10.5880/PIK.2019.018.

Hakkarainen, J., M. E. Szelag, I. Ialongo, C. Retscher, T. Oda, and D. Crisp. 2021. Analyzing nitrogen oxides to carbon dioxide emission ratios from space: A case study of Matimba Power Station in South Africa. *Atmospheric Environment: X* 10:100110. https://doi.org/10.1016/j.aeaoa.2021.100110.

Hale, T. 2020. Transnational actors and transnational governance in global environmental politics. *Annual Review of Political Science* 23(1):203-220. https://doi.org/10.1146/annurev-polisci-050718-032644.

Hamrani, A., A. Akbarzadeh, and C. A. Madramootoo. 2020. Machine learning for predicting greenhouse gas emissions from agricultural soils. *Science of the Total Environment* 741:140338. https://doi.org/10.1016/j.scitotenv.2020.140338.

Han, P., N. Zeng, T. Oda, X. Lin, M. Crippa, D. Guan, G. Janssens-Maenhout, X. Ma, Z. Liu, Y. Shan, S. Tao, H. Wang, R. Wang, L. Wu, X. Yun, Q. Zhang, F. Zhao, and B. Zheng. 2020. Evaluating China's fossil-fuel CO_2 emissions from a comprehensive dataset of nine inventories. *Atmospheric Chemistry and Physics* 20(19):11371-11385. https://doi.org/10.5194/acp-20-11371-2020.

Han, Y., A. Gopal, L. Ouyang, and A. Key. 2021. Estimation of corporate greenhouse gas emissions via machine learning. *arXiv preprint* arXiv:2109.04318. https://doi.org/10.48550/arXiv.2109.04318.

Hansis, E., S. J. Davis, and J. Pongratz. 2015. Relevance of methodological choices for accounting of land use change carbon fluxes. *Global Biogeochemical Cycles* 29(8):1230-1246. https://doi.org/10.1002/2014GB004997.

Harmsen, J. H. M., D. P. van Vuuren, D. R. Nayak, A. F. Hof, L. Höglund-Isaksson, P. L. Lucas, J. B. Nielsen, P. Smith, and E. Stehfest. 2019. Long-term marginal abatement cost curves of non-CO_2 greenhouse gases. *Environmental Science & Policy* 99:136-149. https://doi.org/10.1016/j.envsci.2019.05.013.

Hazan, L., J. Tarniewicz, M. Ramonet, O. Laurent, and A. Abbaris. 2016. Automatic processing of atmospheric CO_2 and CH_4 mole fractions at the ICOS Atmosphere Thematic Centre. *Atmospheric Measurement Techniques* 9(9):4719-4736. https://doi.org/10.5194/amt-9-4719-2016.

He, L., Z.-C. Zeng, T. J. Pongetti, C. Wong, J. Liang, K. R. Gurney, S. Newman, V. Yadav, K. Verhulst, C. E. Miller, R. Duren, C. Frankenberg, P. O. Wennberg, R.-L. Shia, Y. L. Yung, and S. P. Sander. 2019. Atmospheric methane emissions correlate with natural gas consumption from residential and commercial sectors in Los Angeles. *Geophysical Research Letters* 46(14):8563-8571. https://doi.org/https://doi.org/10.1029/2019GL083400.

Heerah, S., I. Frausto-Vicencio, S. Jeong, A. R. Marklein, Y. Ding, A. G. Meyer, H. A. Parker, M. L. Fischer, J. E. Franklin, F. M. Hopkins, and M. Dubey. 2021. Dairy methane emissions in California's San Joaquin Valley inferred with ground-based remote sensing observations in the summer and winter. *Journal of Geophysical Research: Atmospheres* 126(24):e2021JD034785. https://doi.org/https://doi.org/10.1029/2021JD034785.

Heimburger, A. M. F., R. M. Harvey, P. B. Shepson, B. H. Stirm, C. Gore, J. Turnbull, M. O. L. Cambaliza, O. E. Salmon, A.-E. M. Kerlo, T. N. Lavoie, K. J. Davis, T. Lauvaux, A. Karion, C. Sweeney, W. A. Brewer, R. M. Hardesty, and K. R. Gurney. 2017. Assessing the optimized precision of the aircraft mass balance method for measurement of urban greenhouse gas emission rates through averaging. *Elementa: Science of the Anthropocene* 5. https://doi.org/10.1525/elementa.134.

Helfter, C., A. H. Tremper, C. H. Halios, S. Kotthaus, A. Bjorkegren, C. S. B. Grimmond, J. F. Barlow, and E. Nemitz. 2016. Spatial and temporal variability of urban fluxes of methane, carbon monoxide and carbon dioxide above London, UK. *Atmospheric Chemistry and Physics* 16(16):10543-10557. https://doi.org/10.5194/acp-16-10543-2016.

Henne, S., D. Brunner, B. Oney, M. Leuenberger, W. Eugster, I. Bamberger, F. Meinhardt, M. Steinbacher, and L. Emmenegger. 2016. Validation of the Swiss methane emission inventory by atmospheric observations and inverse modelling. *Atmospheric Chemistry and Physics* 16(6):3683-3710. https://doi.org/10.5194/acp-16-3683-2016.

Hennig, R. J., M. Ge, J. Friedrich, K. Lebling, G. Carlock, A. Arcipowska, E. Mangan, H. Biru, A. Tankou, and M. Chaudhury. 2017. *The Data Platform for Climate Research and Action: Introducing Climate Watch*. Presented at American Geophysical Union Fall Meeting, December 1.

Hertwich, E. G., and G. P. Peters. 2009. Carbon footprint of nations: A global, trade-linked analysis. *Environmental Science & Technology* 43(16):6414-6420. https://doi.org/10.1021/es803496a.

Hisano, M., E. B. Searle, and H. Y. H. Chen. 2018. Biodiversity as a solution to mitigate climate change impacts on the functioning of forest ecosystems. *Biological Reviews* 93(1):439-456. https://doi.org/10.1111/brv.12351.

Hmiel, B., V. V. Petrenko, M. N. Dyonisius, C. Buizert, A. M. Smith, P. F. Place, C. Harth, R. Beaudette, Q. Hua, B. Yang, I. Vimont, S. E. Michel, J. P. Severinghaus, C. Etheridge, T. Bromley, J. Schmitt, X. Faïn, R. F. Weiss, and E. Dlugokencky. 2020. Preindustrial $^{14}CH_4$ indicates greater anthropogenic fossil CH_4 emissions. *Nature* 578(7795):409-412. https://doi.org/10.1038/s41586-020-1991-8.

Hoesly, R. M., and S. J. Smith. 2018. Informing energy consumption uncertainty: An analysis of energy data revisions. *Environmental Research Letters* 13(12):124023. https://doi.org/10.1088/1748-9326/aaebc3.

Höglund-Isaksson, L., A. Gómez-Sanabria, Z. Klimont, P. Rafaj, and W. Schöpp. 2020. Technical potentials and costs for reducing global anthropogenic methane emissions in the 2050 timeframe –results from the GAINS model. *Environmental Research Communications* 2(2):025004. https://doi.org/10.1088/2515-7620/ab7457.

Houghton, R. A., and A. A. Nassikas. 2017. Global and regional fluxes of carbon from land use and land cover change 1850–2015. *Global Biogeochemical Cycles* 31(3):456-472. https://doi.org/10.1002/2016GB005546.

Howlett, M. 2009. Policy analytical capacity and evidence-based policy-making: Lessons from Canada. *Canadian Public Administration* 52(2):153-175. https://doi.org/10.1111/j.1754-7121.2009.00070_1.x.

Hsu, A., Y. Cheng, K. Xu, A. Weinfurter, C. Yick, M. Ivanenko, S. Nair, T. Hale, B. Guy, and C. Rosengarten. 2015. Assessing the Wider World of Non-State and Sub-National Climate Action. Yale School of Forestry and Environmental Studies. https://datadrivenlab.org/wp-content/uploads/2020/01/Assessing-the-Wider-World-of-Non-state-and-Sub-national-Climate-Action-Dec10.pdf.

Hsu, A., A. Weinfurter, A. Feierman, Y. Xie, Z. Y. Yeo, K. Lütkehermöller, T. Kuramochi, S. Lui, N. Höhne, and M. Roelfsema. 2018. *Global Climate Action of Regions, States and Businesses.* New Haven, CT: Data Driven Yale. https://datadrivenlab.org/wp-content/uploads/2018/08/YALE-NCI-PBL_Global_climate_action.pdf.

Hsu, A., W. Khoo, N. Goyal, and M. Wainstein. 2020. Next-generation digital ecosystem for climate data mining and knowledge discovery: A review of digital data collection technologies. *Frontiers in Big Data* 3. https://doi.org/10.3389/fdata.2020.00029.

Hsu, A., X. Wang, J. Tan, W. Toh, and N. Goyal. 2022. Predicting European cities' climate mitigation performance using machine learning. https://www.researchsquare.com/article/rs-1450940/v1.

Hsu, Y.-K., T. VanCuren, S. Park, C. Jakober, J. Herner, M. FitzGibbon, D. R. Blake, and D. D. Parrish. 2010. Methane emissions inventory verification in southern California. *Atmospheric Environment* 44(1):1-7. https://doi.org/10.1016/j.atmosenv.2009.10.002.

Humpage, N., H. Boesch, W. Okello, F. Dietrich, J. Chen, M. Lunt, L. Feng, and P. Palmer. 2020. Greenhouse gas column observations from a portable spectrometer in tropical Africa. Presented at 16th International Workshop on Greenhouse Gas Measurements from Space, June 2-5. http://doi.org/10.13140/RG.2.2.14528.97280.

Huntingford, C., E. S. Jeffers, M. B. Bonsall, H. M. Christensen, T. Lees, and H. Yang. 2019. Machine learning and artificial intelligence to aid climate change research and preparedness. *Environmental Research Letters* 14(12):124007. https://doi.org/10.1088/1748-9326/ab4e55.

Hurtt, G. C., A. Andrews, K. Bowman, M. E. Brown, A. Chatterjee, V. Escobar, L. Fatoyinbo, P. Griffith, M. Guy, S. P. Healey, D. J. Jacob, R. Kennedy, S. Lohrenz, M. E. McGroddy, V. Morales, T. Nehrkorn, L. Ott, S. Saatchi, E. Sepulveda Carlo, S. P. Serbin, and H. Tian. 2022. The NASA Carbon Monitoring System Phase 2 synthesis: Scope, findings, gaps and recommended next steps. *Environmental Research Letters* 17(6):063010. https://doi.org/10.1088/1748-9326/ac7407.

Hutchins, M. G., J. D. Colby, G. Marland, and E. Marland. 2017. A comparison of five high-resolution spatially-explicit, fossil-fuel, carbon dioxide emission inventories for the United States. *Mitigation and Adaptation Strategies for Global Change* 22(6):947-972. https://doi.org/10.1007/s11027-016-9709-9.

Hutyra, L. R., R. Duren, K. R. Gurney, N. Grimm, E. A. Kort, E. Larson, and G. Shrestha. 2014. Urbanization and the carbon cycle: Current capabilities and research outlook from the natural sciences perspective. *Earth's Future* 2(10):473-495. https://doi.org/https://doi.org/10.1002/2014EF000255.

Hyperledger. 2021. Decentralized ID and Access Management (DIAM) for IoT Networks. Hyperledger Telecom Special Interest Group. https://www.hyperledger.org/wp-content/uploads/2021/02/HL_LFEdge_WhitePaper_021121_3.pdf.

Iacovidou, E., J. Millward-Hopkins, J. Busch, P. Purnell, C. A. Velis, J. N. Hahladakis, O. Zwirner, and A. Brown. 2017. A pathway to circular economy: Developing a conceptual framework for complex value assessment of resources recovered from waste. *Journal of Cleaner Production* 168:1279-1288. https://doi.org/10.1016/j.jclepro.2017.09.002.

Ibrahim, N., L. Sugar, D. Hoornweg, and C. Kennedy. 2012. Greenhouse gas emissions from cities: Comparison of international inventory frameworks. *Local Environment* 17(2):223-241. https://doi.org/10.1080/13549839.2012.660909.

Idso, C. D., S. B. Idso, and R. C. Balling. 2001. An intensive two-week study of an urban CO_2 dome in Phoenix, Arizona, USA. *Atmospheric Environment* 35(6):995-1000. https://doi.org/https://doi.org/10.1016/S1352-2310(00)00412-X.

IEA (International Energy Agency). 2021a. Global Energy Review: CO_2 Emissions in 2021. https://www.iea.org/data-and-statistics/data-product/global-energy-review-co2-emissions-in-2021.

IEA. 2021b. Emissions Factors 2021. https://www.iea.org/data-and-statistics/data-product/emissions-factors-2021.

IEA. 2021. Key World Energy Statistics 2021c. https://iea.blob.core.windows.net/assets/52f66a88-0b63-4ad2-94a5-29d36e864b82/KeyWorldEnergyStatistics2021.pdf.

Imasu, R., and Y. Tanabe. 2018. Diurnal and seasonal variations of carbon dioxide (CO_2) concentration in urban, suburban, and rural areas around Tokyo. *Atmosphere* 9(10):367.

Ionov, D. V., M. V. Makarova, F. Hase, S. C. Foka, V. S. Kostsov, C. Alberti, T. Blumenstock, T. Warneke, and Y. A. Virolainen. 2021. The CO_2 integral emission by the megacity of St. Petersburg as quantified from ground-based FTIR measurements combined with dispersion modelling. *Atmospheric Chemistry and Physics* 21(14):10939-10963. https://doi.org/10.5194/acp-21-10939-2021.

IPCC (Intergovernmental Panel on Climate Change). 2000. *Land Use, Land Use Change, and Forestry.* R. T. Watson, I. R. Noble, B. Bolin, N. H. Ravindranath, D. J. Verardo, and D. J. Dokken, eds. Cambridge, UK: Cambridge University Press. https://www.ipcc.ch/report/land-use-land-use-change-and-forestry.

IPCC. 2006. *2006 IPCC Guidelines for National Greenhouse Gas Inventories*. S. Eggelston, L. Buendia, K. Miwa, T. Ngara, and K. Tanabe, eds. Hayama, Kanagawa, Japan: Institute for Global Environmental Strategies. https://www.ipcc-nggip.iges.or.jp/public/2006gl/index.html.

IPCC. 2019a. *2019 Refinement to the 2006 IPCC Guidelines for National Greenhouse Gas Inventories*. E. C. Buendia, K. Tanabe, A. Kranjc, B. Jamsranjav, M. Fukuda, S. Ngarize, A. Osako, Y. Pyrozhenko, P. Shermanau and S. Federici, eds. Hayama, Kanagawa, Japan: Institute for Global Environmental Strategies. https://www.ipcc-nggip.iges.or.jp/public/2019rf/index.html.

IPCC. 2019b. *Climate Change and Land: An IPCC Special Report on Climate Change, Desertification, Land Degradation, Sustainable Land Management, Food Security, and Greenhouse Gas Fluxes in Terrestrial Ecosystems*. P. R. Shukla, J. Skea, E. Calvo Buendia, V. Masson-Delmotte, H. -O. Pörtner, D. C. Roberts, P. Zhai, R. Slade, S. Connors, R. van Diemen, M. Ferrat, E. Haughey, S. Luz, S. Neogi, M. Pathak, J. Petzold, J. Portugal Pereira, P. Vyas, E. Huntley, K. Kissick, M. Belkacemi, and J. Malley, eds. Geneva, Switzerland: IPCC. https://www.ipcc.ch/srccl/.

IPCC. 2021. Summary for policymakers. In *Climate Change 2021: The Physical Science Basis. Contribution of Working Group I to the Sixth Assessment Report of the Intergovernmental Panel on Climate Change*. V. Masson-Delmotte, P. Zhai, A. Pirani, S. L. Connors, C. Péan, S. Berger, N. Caud, Y. Chen, L. Goldfarb, M. I. Gomis, M., Huang, K. Leitzell, E. Lonnoy, J. B. R. Matthews, T. K. Maycock, T. Waterfield, O. Yelekçi, R. Yu, and B. Zhou, eds. Cambridge, UK and New York: Cambridge University Press.

IPCC. 2022a. *Climate Change 2022: Impacts, Adaptation and Vulnerability. Contribution of Working Group II to the Sixth Assessment Report of the Intergovernmental Panel on Climate Change*. H.-O. Pörtner, D.C. Roberts, M. Tignor, E. S. Poloczanska, K. Mintenbeck, A. Alegría, M. Craig, S. Langsdorf, S. Löschke, V. Möller, A. Okem, and B. Rama, eds. Cambridge, UK and New York, NY, USA: Cambridge University Press. http://doi.org/10.1017/9781009325844.

IPCC. 2022b. Summary for policymakers. In *Climate Change 2022: Mitigation of Climate Change. Contribution of Working Group III to the Sixth Assessment Report of the Intergovernmental Panel on Climate Change*. P. R. Shukla, J. Skea, R. Slade, A. A. Khourdajie, R. v. Diemen, D. McCollum, M. Pathak, S. Some, P. Vyas, R. Fradera, M. Belkacemi, A. Hasija, G. Lisboa, S. Luz, and J. Malley, eds. Cambridge, UK and New York, NY, USA: Cambridge University Press. http://doi.org/10.1017/9781009157926.001.

IPCC. 2022c. Urban systems and other settlements. In *Climate Change 2022: Mitigation of Climate Change. Contribution of Working Group III to the Sixth Assessment Report of the Intergovernmental Panel on Climate Change*. P. R. Shukla, J. Skea, R. Slade, A. A. Khourdajie, R. v. Diemen, D. McCollum, M. Pathak, S. Some, P. Vyas, R. Fradera, M. Belkacemi, A. Hasija, G. Lisboa, S. Luz, and J. Malley, eds. Cambridge, UK and New York, NY, USA: Cambridge University Press.

ISO (International Organization for Standardization). 2018. ISO 14064-1:2018(en). Greenhouse Gases—Part 1: Specification with Guidance at the Organization Level for Quantification and Reporting of Greenhouse Gas Emissions and Removals. https://www.iso.org/obp/ui#iso:std:iso:14064:-1:ed-2:v1:en.

Ivanova, D., G. Vita, K. Steen-Olsen, K. Stadler, P. C. Melo, R. Wood, and E. G. Hertwich. 2017. Mapping the carbon footprint of EU regions. *Environmental Research Letters* 12(5):054013. https://doi.org/10.1088/1748-9326/aa6da9.

Jackson, R. B., P. Friedlingstein, C. Le Quéré, S. Abernethy, R. M. Andrew, J. G. Canadell, P. Ciais, S. J. Davis, Z. Deng, Z. Liu, J. I. Korsbakken, and G. P. Peters. 2022. Global fossil carbon emissions rebound near pre-COVID-19 levels. *Environmental Research Letters* 17(3):031001. https://doi.org/10.1088/1748-9326/ac55b6.

Jacob, D. J., D. J. Varon, D. H. Cusworth, P. E. Dennison, C. Frankenberg, R. Gautam, L. Guanter, J. Kelley, J. McKeever, L. E. Ott, B. Poulter, Z. Qu, A. K. Thorpe, J. R. Worden, and R. M. Duren. 2022. Quantifying methane emissions from the global scale down to point sources using satellite observations of atmospheric methane. *Atmospheric Chemistry and Physics* 22(14):9617-9646. https://doi.org/10.5194/acp-22-9617-2022.

Janssens-Maenhout, G., B. Pinty, M. Dowell, H. Zunker, E. Andersson, G. Balsamo, J. L. Bézy, T. Brunhes, H. Bösch, B. Bojkov, D. Brunner, M. Buchwitz, D. Crisp, P. Ciais, P. Counet, D. Dee, H. Denier van der Gon, H. Dolman, M. R. Drinkwater, O. Dubovik, R. Engelen, T. Fehr, V. Fernandez, M. Heimann, K. Holmlund, S. Houweling, R. Husband, O. Juvyns, A. Kentarchos, J. Landgraf, R. Lang, A. Löscher, J. Marshall, Y. Meijer, M. Nakajima, P. I. Palmer, P. Peylin, P. Rayner, M. Scholze, B. Sierk, J. Tamminen, and P. Veefkind. 2020. Toward an operational anthropogenic CO_2 emissions monitoring and verification support capacity. *Bulletin of the American Meteorological Society* 101(8):E1439-E1451. https://doi.org/10.1175/BAMS-D-19-0017.1.

Järvi, L., A. Nordbo, H. Junninen, A. Riikonen, J. Moilanen, E. Nikinmaa, and T. Vesala. 2012. Seasonal and annual variation of carbon dioxide surface fluxes in Helsinki, Finland, in 2006–2010. *Atmospheric Chemistry and Physics* 12(18):8475-8489. https://doi.org/10.5194/acp-12-8475-2012.

Jervis, D., J. McKeever, B. O. A. Durak, J. J. Sloan, D. Gains, D. J. Varon, A. Ramier, M. Strupler, and E. Tarrant. 2021. The GHGSat-D imaging spectrometer. *Atmospheric Measurement Techniques* 14(3):2127-2140. https://doi.org/10.5194/amt-14-2127-2021.

Jones, C. M., and D. M. Kammen. 2015. A Consumption-Based Greenhouse Gas Inventory of San Francisco Bay Area Neighborhoods, Cities and Counties: Prioritizing Climate Action for Different Locations. https://escholarship.org/uc/item/2sn7m83z.

Jones, M. W., R. M. Andrew, G. P. Peters, G. Janssens-Maenhout, A. J. De-Gol, P. Ciais, P. K. Patra, F. Chevallier, and C. Le Quéré. 2021. Gridded fossil CO_2 emissions and related O_2 combustion consistent with national inventories 1959–2018. Scientific Data 8(1):2. https://doi.org/10.1038/s41597-020-00779-6.

Jongaramrungruang, S., C. Frankenberg, G. Matheou, A. K. Thorpe, D. R. Thompson, L. Kuai, and R. M. Duren. 2019. Towards accurate methane point-source quantification from high-resolution 2-D plume imagery. *Atmospheric Measurement Techniques* 12(12):6667-6681. https://doi.org/10.5194/amt-12-6667-2019.

Kaack, L. H., P. L. Donti, E. Strubell, G. Kamiya, F. Creutzig, and D. Rolnick. 2022. Aligning artificial intelligence with climate change mitigation. *Nature Climate Change* 12(6):518-527. https://doi.org/10.1038/s41558-022-01377-7.

Kaminski, T., M. Scholze, P. Rayner, M. Voßbeck, M. Buchwitz, M. Reuter, W. Knorr, H. Chen, A. Agustí-Panareda, A. Löscher, and Y. Meijer. 2022. Assimilation of atmospheric CO_2 observations from space can support national CO_2 emission inventories. *Environmental Research Letters* 17(1):014015. https://doi.org/10.1088/1748-9326/ac3cea.

Kanemoto, K., D. Moran, and E. G. Hertwich. 2016. Mapping the carbon footprint of nations. *Environmental Science & Technology* 50(19):10512-10517. https://doi.org/10.1021/acs.est.6b03227.

Karion, A., W. Callahan, M. Stock, S. Prinzivalli, K. R. Verhulst, J. Kim, P. K. Salameh, I. Lopez-Coto, and J. Whetstone. 2020. Greenhouse gas observations from the Northeast Corridor tower network. *Earth System Science Data* 12(1):699-717. https://doi.org/10.5194/essd-12-699-2020.

Kawanishi, M., and R. Fujikura. 2018. Evaluation of enabling factors for sustainable national greenhouse gas inventory in developing countries. *International Journal of Environmental Science and Development* 9(10):290-297. https://doi.org/10.18178/ijesd.2018.9.10.1116.

Keeling, C. D. 1973. Industrial production of carbon dioxide from fossil fuels and limestone. *Tellus* 25(2):174-198. https://doi.org/10.1111/j.2153-3490.1973.tb01604.x.

Keeling, C. D., R. B. Bacastow, A. E. Bainbridge, C. A. Ekdahl Jr., P. R. Guenther, L. S. Waterman, and J. F. S. Chin. 1976. Atmospheric carbon dioxide variations at Mauna Loa Observatory, Hawaii. *Tellus* 28(6):538-551. https://doi.org/10.3402/tellusa.v28i6.11322.

Keeling, R. F., S. C. Piper, and M. Heimann. 1996. Global and hemispheric CO_2 sinks deduced from changes in atmospheric O_2 concentration. *Nature* 381(6579):218-221. https://doi.org/10.1038/381218a0.

Keyes, T., G. Ridge, M. Klein, N. Phillips, R. Ackley, and Y. Yang. 2020. An enhanced procedure for urban mobile methane leak detection. *Heliyon* 6(10):e04876. https://doi.org/10.1016/j.heliyon.2020.e04876.

Kim, D. G., B. Bond-Lamberty, Y. Ryu, B. Seo, and D. Papale. 2022a. Ideas and perspectives: Enhancing research and monitoring of carbon pools and land-to-atmosphere greenhouse gases exchange in developing countries. *Biogeosciences* 19(5):1435-1450. https://doi.org/10.5194/bg-19-1435-2022.

Kim, J., A. A. Shusterman, K. J. Lieschke, C. Newman, and R. C. Cohen. 2018. The BErkeley Atmospheric CO_2 Observation Network: Field calibration and evaluation of low-cost air quality sensors. *Atmospheric Measurement Techniques* 11(4):1937-1946. https://doi.org/10.5194/amt-11-1937-2018.

Kim, J., A. J. Turner, H. L. Fitzmaurice, E. R. Delaria, C. Newman, P. J. Wooldridge, and R. C. Cohen. 2022b. Observing annual trends in vehicular CO_2 emissions. *Environmental Science & Technology* 56(7):3925-3931. https://doi.org/10.1021/acs.est.1c06828.

Klaaßen, L., and C. Stoll. 2021. Harmonizing corporate carbon footprints. *Nature Communications* 12(1):6149. https://doi.org/10.1038/s41467-021-26349-x.

Klimont, Z., K. Kupiainen, C. Heyes, P. Purohit, J. Cofala, P. Rafaj, J. Borken-Kleefeld, and W. Schöpp. 2017. Global anthropogenic emissions of particulate matter including black carbon. *Atmospheric Chemistry and Physics* 17(14):8681-8723. https://doi.org/10.5194/acp-17-8681-2017.

Kljun, N., P. Calanca, M. W. Rotach, and H. P. Schmid. 2015. A simple two-dimensional parameterisation for Flux Footprint Prediction (FFP). *Geoscientific Model Development* 8(11):3695-3713. https://doi.org/10.5194/gmd-8-3695-2015.

Kloppenburg, S., A. Gupta, S. R. L. Kruk, S. Makris, R. Bergsvik, P. Korenhof, H. Solman, and H. M. Toonen. 2022. Scrutinizing environmental governance in a digital age: New ways of seeing, participating, and intervening. *One Earth* 5(3):232-241. https://doi.org/10.1016/j.oneear.2022.02.004.

Knapp, M., B. Hemmer, R. Kleinschek, M. Sindram, T. Schmitt, L. Pilz, B. Burger, and A. Butz. 2022. Towards carbon dioxide emission estimation with a stationary hyperspectral camera. Presented at EGU General Assembly 2022, Vienna, Austria & Online, May 23-27. https://doi.org/10.5194/egusphere-egu22-3924.

Kona, A., P. Bertoldi, F. Monforti-Ferrario, S. Rivas, and J. F. Dallemand. 2018. Covenant of mayors signatories leading the way towards 1.5 degree global warming pathway. *Sustainable Cities and Society* 41:568-575. https://doi.org/10.1016/j.scs.2018.05.017.

Kona, A., F. Monforti-Ferrario, P. Bertoldi, M. G. Baldi, G. Kakoulaki, N. Vetters, C. Thiel, G. Melica, E. Lo Vullo, A. Sgobbi, C. Ahlgren, and B. Posnic. 2021. Global Covenant of Mayors, a dataset of greenhouse gas emissions for 6200 cities in Europe and the Southern Mediterranean countries. *Earth System Science Data* 13(7):3551-3564. https://doi.org/10.5194/essd-13-3551-2021.

Kováč, D., A. Ač, L. Šigut, J. Peñuelas, J. Grace, and O. Urban. 2022. Combining NDVI, PRI and the quantum yield of solar-induced fluorescence improves estimations of carbon fluxes in deciduous and evergreen forests. *Science of the Total Environment* 829:154681. https://doi.org/10.1016/j.scitotenv.2022.154681.

Krautwurst, S., K. Gerilowski, J. Borchardt, N. Wildmann, M. Gałkowski, J. Swolkień, J. Marshall, A. Fiehn, A. Roiger, T. Ruhtz, C. Gerbig, J. Necki, J. P. Burrows, A. Fix, and H. Bovensmann. 2021. Quantification of CH_4 coal mining emissions in Upper Silesia by passive airborne remote sensing observations with the Methane Airborne MAPper (MAMAP) instrument during the CO_2 and Methane (CoMet) campaign. *Atmospheric Chemistry and Physics* 21(23):17345-17371. https://doi.org/10.5194/acp-21-17345-2021.

Krings, T., B. Neininger, K. Gerilowski, S. Krautwurst, M. Buchwitz, J. P. Burrows, C. Lindemann, T. Ruhtz, D. Schüttemeyer, and H. Bovensmann. 2018. Airborne remote sensing and in situ measurements of atmospheric CO_2 to quantify point source emissions. *Atmospheric Measurement Techniques* 11(2):721-739. https://doi.org/10.5194/amt-11-721-2018.

Lamb, B. K., S. L. Edburg, T. W. Ferrara, T. Howard, M. R. Harrison, C. E. Kolb, A. Townsend-Small, W. Dyck, A. Possolo, and J. R. Whetstone. 2015. Direct measurements show decreasing methane emissions from natural gas local distribution systems in the United States. *Environmental Science & Technology* 49(8):5161-5169. https://doi.org/10.1021/es505116p.

Lauvaux, T., N. L. Miles, A. Deng, S. J. Richardson, M. O. Cambaliza, K. J. Davis, B. Gaudet, K. R. Gurney, J. Huang, D. O'Keefe, Y. Song, A. Karion, T. Oda, R. Patarasuk, I. Razlivanov, D. Sarmiento, P. Shepson, C. Sweeney, J. Turnbull, and K. Wu. 2016. High-resolution atmospheric inversion of urban CO_2 emissions during the dormant season of the Indianapolis Flux Experiment (INFLUX). *Journal of Geophysical Research: Atmospheres* 121(10):5213-5236. https://doi.org/10.1002/2015JD024473.

Lauvaux, T., K. R. Gurney, N. L. Miles, K. J. Davis, S. J. Richardson, A. Deng, B. J. Nathan, T. Oda, J. A. Wang, L. Hutyra, and J. Turnbull. 2020. Policy-relevant assessment of urban CO_2 emissions. *Environmental Science & Technology* 54(16):10237-10245. https://doi.org/10.1021/acs.est.0c00343.

Lauvaux, T., C. Giron, M. Mazzolini, A. d'Aspremont, R. Duren, D. Cusworth, D. Shindell, and P. Ciais. 2022. Global assessment of oil and gas methane ultra-emitters. *Science* 375(6580):557-561. https://doi.org/10.1126/science.abj4351.

Le Quéré, C., R. B. Jackson, M. W. Jones, A. J. P. Smith, S. Abernethy, R. M. Andrew, A. J. De-Gol, D. R. Willis, Y. Shan, J. G. Canadell, P. Friedlingstein, F. Creutzig, and G. P. Peters. 2020. Temporary reduction in daily global CO_2 emissions during the COVID-19 forced confinement. *Nature Climate Change* 10(7):647-653. https://doi.org/10.1038/s41558-020-0797-x.

Lee, D. J., and B. Stvilia. 2014. Developing a data identifier taxonomy. *Cataloging & Classification Quarterly* 52(3):303-336. https://doi.org/10.1080/01639374.2014.880166.

Levy, B. S., and J. A. Patz. 2015. Climate change, human rights, and social justice. *Annals of Global Health* 81(3):310-322. https://doi.org/10.1016/j.aogh.2015.08.008.

Lin, J. C., L. Mitchell, E. Crosman, D. L. Mendoza, M. Buchert, R. Bares, B. Fasoli, D. R. Bowling, D. Pataki, D. Catharine, C. Strong, K. R. Gurney, R. Patarasuk, M. Baasandorj, A. Jacques, S. Hoch, J. Horel, and J. Ehleringer. 2018. CO_2 and carbon emissions from cities: Linkages to air quality, socioeconomic activity, and stakeholders in the Salt Lake City urban area. *Bulletin of the American Meteorological Society* 99(11):2325-2339. https://doi.org/10.1175/BAMS-D-17-0037.1.

Linden, O., A. Jerneloev, and J. Egerup. 2004. *The Environmental Impacts of the Gulf War 1991*. IIASA Interim Report. Laxenburg, Austria: International Institute for Applied Systems Analysis. https://pure.iiasa.ac.at/id/eprint/7427/.

Lindroth, A., and L. Tranvik. 2021. Accounting for all territorial emissions and sinks is important for development of climate mitigation policies. *Carbon Balance and Management* 16(1):10. https://doi.org/10.1186/s13021-021-00173-8.

Liu, Z., D. Guan, W. Wei, S. J. Davis, P. Ciais, J. Bai, S. Peng, Q. Zhang, K. Hubacek, G. Marland, R. J. Andres, D. Crawford-Brown, J. Lin, H. Zhao, C. Hong, T. A. Boden, K. Feng, G. P. Peters, F. Xi, J. Liu, Y. Li, Y. Zhao, N. Zeng, and K. He. 2015. Reduced carbon emission estimates from fossil fuel combustion and cement production in China. *Nature* 524(7565):335-338. https://doi.org/10.1038/nature14677.

Liu, Z., P. Ciais, Z. Deng, S. J. Davis, B. Zheng, Y. Wang, D. Cui, B. Zhu, X. Dou, P. Ke, T. Sun, R. Guo, H. Zhong, O. Boucher, F.-M. Bréon, C. Lu, R. Guo, J. Xue, E. Boucher, K. Tanaka, and F. Chevallier. 2020a. Carbon Monitor, a near-real-time daily dataset of global CO_2 emission from fossil fuel and cement production. *Scientific Data* 7(1):392. https://doi.org/10.1038/s41597-020-00708-7.

Liu, Z., P. Ciais, Z. Deng, R. Lei, S. J. Davis, S. Feng, B. Zheng, D. Cui, X. Dou, B. Zhu, R. Guo, P. Ke, T. Sun, C. Lu, P. He, Y. Wang, X. Yue, Y. Wang, Y. Lei, H. Zhou, Z. Cai, Y. Wu, R. Guo, T. Han, J. Xue, O. Boucher, E. Boucher, F. Chevallier, K. Tanaka, Y. Wei, H. Zhong, C. Kang, N. Zhang, B. Chen, F. Xi, M. Liu, F.-M. Bréon, Y. Lu, Q. Zhang, D. Guan, P. Gong, D. M. Kammen, K. He, and H. J. Schellnhuber. 2020b. Near-real-time monitoring of global CO_2 emissions reveals the effects of the COVID-19 pandemic. *Nature Communications* 11(1):5172. https://doi.org/10.1038/s41467-020-18922-7.

Lopez, M., M. Schmidt, M. Delmotte, A. Colomb, V. Gros, C. Janssen, S. J. Lehman, D. Mondelain, O. Perrussel, M. Ramonet, I. Xueref-Remy, and P. Bousquet. 2013. CO, NO_x and $^{13}CO_2$ as tracers for fossil fuel CO_2: Results from a pilot study in Paris during winter 2010. *Atmospheric Chemistry and Physics* 13(15):7343-7358. https://doi.org/10.5194/acp-13-7343-2013.

López-Ballesteros, A., J. Beck, A. Bombelli, E. Grieco, E. K. Lorencová, L. Merbold, C. Brümmer, W. Hugo, R. Scholes, D. Vačkář, A. Vermeulen, M. Acosta, K. Butterbach-Bahl, J. Helmschrot, D.-G. Kim, M. Jones, V. Jorch, M. Pavelka, I. Skjelvan, and M. Saunders. 2018. Towards a feasible and representative pan-African research infrastructure network for GHG observations. *Environmental Research Letters* 13(8):085003. https://doi.org/10.1088/1748-9326/aad66c.

Lopez-Coto, I., X. Ren, O. E. Salmon, A. Karion, P. B. Shepson, R. R. Dickerson, A. Stein, K. Prasad, and J. R. Whetstone. 2020. Wintertime CO_2, CH_4, and CO emissions estimation for the Washington, DC–Baltimore metropolitan area using an inverse modeling technique. *Environmental Science & Technology* 54(5):2606-2614. https://doi.org/10.1021/acs.est.9b06619.

Lu, X., D. J. Jacob, H. Wang, J. D. Maasakkers, Y. Zhang, T. R. Scarpelli, L. Shen, Z. Qu, M. P. Sulprizio, H. Nesser, A. A. Bloom, S. Ma, J. R. Worden, S. Fan, R. J. Parker, H. Boesch, R. Gautam, D. Gordon, M. D. Moran, F. Reuland, C. A. O. Villasana, and A. Andrews. 2022. Methane emissions in the United States, Canada, and Mexico: Evaluation of national methane emission inventories and 2010–2017 sectoral trends by inverse analysis of in situ (GLOBALVIEWplus CH_4 ObsPack) and satellite (GOSAT) atmospheric observations. *Atmospheric Chemistry and Physics* 22(1):395-418. https://doi.org/10.5194/acp-22-395-2022.

Lux, Z. A., D. Thatmann, S. Zickau, and F. Beierle. 2020. Distributed-ledger-based authentication with decentralized identifiers and verifiable credentials. Presented at 2020 2nd Conference on Blockchain Research & Applications for Innovative Networks and Services (BRAINS), September 28-30.

Lyon, D. R., B. Hmiel, R. Gautam, M. Omara, K. A. Roberts, Z. R. Barkley, K. J. Davis, N. L. Miles, V. C. Monteiro, S. J. Richardson, S. Conley, M. L. Smith, D. J. Jacob, L. Shen, D. J. Varon, A. Deng, X. Rudelis, N. Sharma, K. T. Story, A. R. Brandt, M. Kang, E. A. Kort, A. J. Marchese, and S. P. Hamburg. 2021. Concurrent variation in oil and gas methane emissions and oil price during the COVID-19 pandemic. *Atmospheric Chemistry and Physics* 21(9):6605-6626. https://doi.org/10.5194/acp-21-6605-2021.

Maasakkers, J. D., D. J. Jacob, M. P. Sulprizio, A. J. Turner, M. Weitz, T. Wirth, C. Hight, M. DeFigueiredo, M. Desai, R. Schmeltz, L. Hockstad, A. A. Bloom, K. W. Bowman, S. Jeong, and M. L. Fischer. 2016. Gridded national inventory of U.S. methane emissions. *Environmental Science & Technology* 50(23):13123-13133. https://doi.org/10.1021/acs.est.6b02878.

MacBean, N., F. Maignan, C. Bacour, P. Lewis, P. Peylin, L. Guanter, P. Köhler, J. Gómez-Dans, and M. Disney. 2018. Strong constraint on modelled global carbon uptake using solar-induced chlorophyll fluorescence data. *Scientific Reports* 8(1):1973. https://doi.org/10.1038/s41598-018-20024-w.

Machida, T., H. Matsueda, Y. Sawa, Y. Nakagawa, K. Hirotani, N. Kondo, K. Goto, T. Nakazawa, K. Ishikawa, and T. Ogawa. 2008. Worldwide measurements of atmospheric CO_2 and other trace gas species using commercial airlines. *Journal of Atmospheric and Oceanic Technology* 25(10):1744-1754. https://doi.org/10.1175/2008JTECHA1082.1.

Mallia, D. V., L. E. Mitchell, L. Kunik, B. Fasoli, R. Bares, K. R. Gurney, D. L. Mendoza, and J. C. Lin. 2020. Constraining urban CO_2 emissions using mobile observations from a light rail public transit platform. *Environmental Science & Technology* 54(24):15613-15621. https://doi.org/10.1021/acs.est.0c04388.

Manning, A. J. 2011. The challenge of estimating regional trace gas emissions from atmospheric observations. *Philosophical Transactions of the Royal Society A: Mathematical, Physical and Engineering Sciences* 369(1943):1943-1954. https://doi.org/10.1098/rsta.2010.0321.

Manning, A. J., S. O'Doherty, A. R. Jones, P. G. Simmonds, and R. G. Derwent. 2011. Estimating UK methane and nitrous oxide emissions from 1990 to 2007 using an inversion modeling approach. *Journal of Geophysical Research: Atmospheres* 116(D2). https://doi.org/10.1029/2010JD014763.

Manning, A. J., A. L. Redington, D. Say, S. O'Doherty, D. Young, P. G. Simmonds, M. K. Vollmer, J. Mühle, J. Arduini, G. Spain, A. Wisher, M. Maione, T. J. Schuck, K. Stanley, S. Reimann, A. Engel, P. B. Krummel, P. J. Fraser, C. M. Harth, P. K. Salameh, R. F. Weiss, R. Gluckman, P. N. Brown, J. D. Watterson, and T. Arnold. 2021. Evidence of a recent decline in UK emissions of hydrofluorocarbons determined by the InTEM inverse model and atmospheric measurements. *Atmospheric Chemistry and Physics* 21(16):12739-12755. https://doi.org/10.5194/acp-21-12739-2021.

Marklein, A. R., D. Meyer, M. L. Fischer, S. Jeong, T. Rafiq, M. Carr, and F. M. Hopkins. 2021. Facility-scale inventory of dairy methane emissions in California: Implications for mitigation. *Earth System Science Data* 13(3):1151-1166. https://doi.org/10.5194/essd-13-1151-2021.

Marland, G. 2008. Uncertainties in accounting for CO_2 from fossil fuels. *Journal of Industrial Ecology* 12(2):136-139. https://doi.org/10.1111/j.1530-9290.2008.00014.x.

Marland, G., and R. M. Rotty. 1984. Carbon dioxide emissions from fossil fuels: A procedure for estimation and results for 1950–1982. *Tellus B* 36B(4):232-261. https://doi.org/10.1111/j.1600-0889.1984.tb00245.x.

Marlowe, J., and A. Clarke. 2022. Carbon accounting: A systematic literature review and directions for future research. *Green Finance* 4(1):71-87. https://doi.org/10.3934/GF.2022004.

Masanet, E., A. Shehabi, N. Lei, S. Smith, and J. Koomey. 2020. Recalibrating global data center energy-use estimates. *Science* 367(6481):984-986. https://doi.org/10.1126/science.aba3758.

Matsueda, H., H. Y. Inoue, and M. Ishii. 2002. Aircraft observation of carbon dioxide at 8–13 km altitude over the western Pacific from 1993 to 1999. *Tellus B* 54(1):1-21. https://doi.org/10.1034/j.1600-0889.2002.00304.x.

Matsueda, H., T. Machida, Y. Sawa, Y. Nakagawa, K. Hirotani, H. Ikeda, N. Kondo, and K. Goto. 2008. Evaluation of atmospheric CO_2 measurements from new flask air sampling of JAL airliner observations. *Papers in Meteorology and Geophysics* 59:1-17. https://doi.org/10.2467/mripapers.59.1.

McGlynn, E., S. Li, M. F. Berger, M. Amend, and K. L. Harper. 2022. Addressing uncertainty and bias in land use, land use change, and forestry greenhouse gas inventories. *Climatic Change* 170(1):5. https://doi.org/10.1007/s10584-021-03254-2.

Meijer, Y. 2022. International satellite monitoring and top-down constraints on greenhouse gas emissions. Presented at Greenhouse Gas Emissions Monitoring, Inventories, and Data Integration: Understanding the Landscape. National Academies of Sciences, Engineering, and Medicine, Board on Atmospheric Sciences and Climate, Washington, DC, June 2.

Meinshausen, M., E. Vogel, A. Nauels, K. Lorbacher, N. Meinshausen, D. M. Etheridge, P. J. Fraser, S. A. Montzka, P. J. Rayner, C. M. Trudinger, P. B. Krummel, U. Beyerle, J. G. Canadell, J. S. Daniel, I. G. Enting, R. M. Law, C. R. Lunder, S. O'Doherty, R. G. Prinn, S. Reimann, M. Rubino, G. J. M. Velders, M. K. Vollmer, R. H. J. Wang, and R. Weiss. 2017. Historical greenhouse gas concentrations for climate modelling (CMIP6). *Geoscientific Model Development* 10(5):2057-2116. https://doi.org/10.5194/gmd-10-2057-2017.

Meng, T., X. Jing, Z. Yan, and W. Pedrycz. 2020. A survey on machine learning for data fusion. *Information Fusion* 57:115-129. https://doi.org/10.1016/j.inffus.2019.12.001.

Miles, N. L., S. J. Richardson, T. Lauvaux, K. J. Davis, N. V. Balashov, A. Deng, J. C. Turnbull, C. Sweeney, K. R. Gurney, R. Patarasuk, I. Razlivanov, M. O. L. Cambaliza, and P. B. Shepson. 2017. Quantification of urban atmospheric boundary layer greenhouse gas dry mole fraction enhancements in the dormant season: Results from the Indianapolis Flux Experiment (INFLUX). *Elementa: Science of the Anthropocene* 5. https://doi.org/10.1525/elementa.127.

Miller, J. B., S. J. Lehman, S. A. Montzka, C. Sweeney, B. R. Miller, A. Karion, C. Wolak, E. J. Dlugokencky, J. Southon, J. C. Turnbull, and P. P. Tans. 2012. Linking emissions of fossil fuel CO_2 and other anthropogenic trace gases using atmospheric $^{14}CO_2$. *Journal of Geophysical Research: Atmospheres* 117(D8). https://doi.org/10.1029/2011JD017048.

Miller, J. B., S. J. Lehman, K. R. Verhulst, C. E. Miller, R. M. Duren, V. Yadav, S. Newman, and C. D. Sloop. 2020. Large and seasonally varying biospheric CO_2 fluxes in the Los Angeles megacity revealed by atmospheric radiocarbon. *Proceedings of the National Academy of Sciences* 117(43):26681-26687. https://doi.org/10.1073/pnas.2005253117.

Miller, S. M., A. M. Michalak, R. G. Detmers, O. P. Hasekamp, L. M. P. Bruhwiler, and S. Schwietzke. 2019. China's coal mine methane regulations have not curbed growing emissions. *Nature Communications* 10(1):303. https://doi.org/10.1038/s41467-018-07891-7.

Milojevic-Dupont, N., and F. Creutzig. 2021. Machine learning for geographically differentiated climate change mitigation in urban areas. *Sustainable Cities and Society* 64:102526. https://doi.org/10.1016/j.scs.2020.102526.

Mingle, J. 2019. Methane detectives: Can a wave of new technology slash natural gas leaks? *Yale Environment 360*. https://e360.yale.edu/features/methane-detectives-can-a-wave-of-new-technology-slash-natural-gas-leaks.

Minx, J., G. Baiocchi, T. Wiedmann, J. Barrett, F. Creutzig, K. Feng, M. Förster, P.-P. Pichler, H. Weisz, and K. Hubacek. 2013. Carbon footprints of cities and other human settlements in the UK. *Environmental Research Letters* 8(3):035039. https://doi.org/10.1088/1748-9326/8/3/035039.

Minx, J. C., W. F. Lamb, R. M. Andrew, J. G. Canadell, M. Crippa, N. Döbbeling, P. M. Forster, D. Guizzardi, J. Olivier, G. P. Peters, J. Pongratz, A. Reisinger, M. Rigby, M. Saunois, S. J. Smith, E. Solazzo, and H. Tian. 2021. A comprehensive and synthetic dataset for global, regional, and national greenhouse gas emissions by sector 1970–2018 with an extension to 2019. *Earth System Science Data* 13(11):5213-5252. https://doi.org/10.5194/essd-13-5213-2021.

Mitchell, A. L., D. S. Tkacik, J. R. Roscioli, S. C. Herndon, T. I. Yacovitch, D. M. Martinez, T. L. Vaughn, L. L. Williams, M. R. Sullivan, C. Floerchinger, M. Omara, R. Subramanian, D. Zimmerle, A. J. Marchese, and A. L. Robinson. 2015. Measurements of methane emissions from natural gas gathering facilities and processing plants: Measurement results. *Environmental Science & Technology* 49(5):3219-3227. https://doi.org/10.1021/es5052809.

Mitchell, L. E., E. T. Crosman, A. A. Jacques, B. Fasoli, L. Leclair-Marzolf, J. Horel, D. R. Bowling, J. R. Ehleringer, and J. C. Lin. 2018. Monitoring of greenhouse gases and pollutants across an urban area using a light-rail public transit platform. *Atmospheric Environment* 187:9-23. https://doi.org/10.1016/j.atmosenv.2018.05.044.

Monteil, G., G. Broquet, M. Scholze, M. Lang, U. Karstens, C. Gerbig, F. T. Koch, N. E. Smith, R. L. Thompson, I. T. Luijkx, E. White, A. Meesters, P. Ciais, A. L. Ganesan, A. Manning, M. Mischurow, W. Peters, P. Peylin, J. Tarniewicz, M. Rigby, C. Rödenbeck, A. Vermeulen, and E. M. Walton. 2020. The regional European atmospheric transport inversion comparison, EUROCOM: First results on European-wide terrestrial carbon fluxes for the period 2006–2015. *Atmospheric Chemistry and Physics* 20(20):12063-12091. https://doi.org/10.5194/acp-20-12063-2020.

Montzka, S. A., G. S. Dutton, P. Yu, E. Ray, R. W. Portmann, J. S. Daniel, L. Kuijpers, B. D. Hall, D. Mondeel, C. Siso, J. D. Nance, M. Rigby, A. J. Manning, L. Hu, F. Moore, B. R. Miller, and J. W. Elkins. 2018. An unexpected and persistent increase in global emissions of ozone-depleting CFC-11. *Nature* 557(7705):413-417. https://doi.org/10.1038/s41586-018-0106-2.

Montzka, S. A., G. S. Dutton, R. W. Portmann, M. P. Chipperfield, S. Davis, W. Feng, A. J. Manning, E. Ray, M. Rigby, B. D. Hall, C. Siso, J. D. Nance, P. B. Krummel, J. Mühle, D. Young, S. O'Doherty, P. K. Salameh, C. M. Harth, R. G. Prinn, R. F. Weiss, J. W. Elkins, H. Walter-Terrinoni, and C. Theodoridi. 2021. A decline in global CFC-11 emissions during 2018–2019. *Nature* 590(7846):428-432. https://doi.org/10.1038/s41586-021-03260-5.

Mora, C., R. L. Rollins, K. Taladay, M. B. Kantar, M. K. Chock, M. Shimada, and E. C. Franklin. 2018. Bitcoin emissions alone could push global warming above 2°C. *Nature Climate Change* 8(11):931-933. https://doi.org/10.1038/s41558-018-0321-8.

Moran, D., K. Kanemoto, M. Jiborn, R. Wood, J. Többen, and K. C. Seto. 2018. Carbon footprints of 13 000 cities. *Environmental Research Letters* 13(6):064041. https://doi.org/10.1088/1748-9326/aac72a.

Moran, D., P. P. Pichler, H. Zheng, H. Muri, J. Klenner, D. Kramel, J. Többen, H. Weisz, T. Wiedmann, A. Wyckmans, A. H. Strømman, and K. R. Gurney. 2022. Estimating CO_2 emissions for 108 000 European cities. *Earth System Science Data* 14(2):845-864. https://doi.org/10.5194/essd-14-845-2022.

Morrison, R., N. C. Mazey, and S. C. Wingreen. 2020. The DAO controversy: The case for a new species of corporate governance? *Frontiers in Blockchain* 3(25). https://doi.org/10.3389/fbloc.2020.00025.

Mostafavi Pak, N., S. Ars, B. Lehman, D. Weaver, F. R. Vogel, and D. Wunch. 2019. Methane measurements using portable fourier transform spectrometers in the Greater Toronto Area. Presented at American Geophysical Union Fall Meeting, December 1.

Mueller, K. L., T. Lauvaux, K. R. Gurney, G. Roest, S. Ghosh, S. M. Gourdji, A. Karion, P. DeCola, and J. Whetstone. 2021. An emerging GHG estimation approach can help cities achieve their climate and sustainability goals. *Environmental Research Letters* 16(8):084003. https://doi.org/10.1088/1748-9326/ac0f25.

Müller, M., P. Graf, J. Meyer, A. Pentina, D. Brunner, F. Perez-Cruz, C. Hüglin, and L. Emmenegger. 2020. Integration and calibration of non-dispersive infrared (NDIR) CO_2 low-cost sensors and their operation in a sensor network covering Switzerland. *Atmospheric Measurement Techniques* 13(7):3815-3834. https://doi.org/10.5194/amt-13-3815-2020.

Muralikrishna, I. V., and V. Manickam. 2017. *Environmental Management: Science and Engineering for Industry*. Kidlington, Oxford, UK: Butterworth-Heinemann.

NASEM (National Academies of Sciences, Engineering, and Medicine). 2016. *Attribution of Extreme Weather Events in the Context of Climate Change*. Washington, DC: The National Academies Press. https://doi.org/10.17226/21852.

NASEM. 2018. *Improving Characterization of Anthropogenic Methane Emissions in the United States*. Washington, DC: The National Academies Press.

NASEM. 2019. *Reproducibility and Replicability in Science*. Washington, DC: The National Academies Press.

Nassar, R., T. G. Hill, C. A. McLinden, D. Wunch, D. B. A. Jones, and D. Crisp. 2017. Quantifying CO_2 emissions from individual power plants from space. *Geophysical Research Letters* 44(19):10045-10053. https://doi.org/10.1002/2017GL074702.

Nathan, B. J., T. Lauvaux, J. C. Turnbull, S. J. Richardson, N. L. Miles, and K. R. Gurney. 2018. Source sector attribution of CO_2 emissions using an urban CO/CO_2 bayesian inversion system. *Journal of Geophysical Research: Atmospheres* 123(23):13611-13621. https://doi.org/10.1029/2018JD029231.

NDRC (National Development and Reform Commission). 2004. *The People's Republic of China's initial national communication on climate change*. Beijing: China Planning Press.

NDRC. 2012. *The People's Republic of China second national communication on climate change*. Beijing: China Planning Press.

NDRC. 2016. *The People's Republic of China first biennial update report on climate change of China*. Beijing: China Planning Press.

NDRC. 2019a. *The People's Republic of China second biennial update report on climate change of China*. Beijing: China Planning Press.

NDRC. 2019b. *The People's Republic of China third national communication on climate change*. Beijing: China Planning Press.

Nicholls, D., F. Barnes, F. Acrea, C. Chen, L. Y. Buluç, and M. M. Parker. 2015. *Top-down and bottom-up approaches to greenhouse gas inventory methods—a comparison between national- and forest-scale reporting methods*. Gen. Tech. Rep. PNW-GTR-906. Portland, OR: U.S. Department of Agriculture, Forest Service, Pacific Northwest Research Station.

Nicolini, G., G. Antoniella, F. Carotenuto, A. Christen, P. Ciais, C. Feigenwinter, B. Gioli, S. Stagakis, E. Velasco, R. Vogt, H. C. Ward, J. Barlow, N. Chrysoulakis, P. Duce, M. Graus, C. Helfter, B. Heusinkveld, L. Järvi, T. Karl, S. Marras, V. Masson, B. Matthews, F. Meier, E. Nemitz, S. Sabbatini, D. Scherer, H. Schume, C. Sirca, G.-J. Steeneveld, C. Vagnoli, Y. Wang, A. Zaldei, B. Zheng, and D. Papale. 2022. Direct observations of CO_2 emission reductions due to COVID-19 lockdown across European urban districts. *Science of the Total Environment* 830:154662. https://doi.org/10.1016/j.scitotenv.2022.154662.

NISO (National Information Standards Organization). 2013. NISO RP-15-2013, Recommended Practices for Online Supplemental Journal Article Materials. https://www.niso.org/publications/niso-rp-15-2013-recommended-practices-online-supplemental-journal-article-materials.

Niu, D., K. Wang, J. Wu, L. Sun, Y. Liang, X. Xu, and X. Yang. 2020. Can China achieve its 2030 carbon emissions commitment? Scenario analysis based on an improved general regression neural network. *Journal of Cleaner Production* 243:118558. https://doi.org/10.1016/j.jclepro.2019.118558.

Nosek, B. A., G. Alter, G. C. Banks, D. Borsboom, S. D. Bowman, S. J. Breckler, S. Buck, C. D. Chambers, G. Chin, G. Christensen, M. Contestabile, A. Dafoe, E. Eich, J. Freese, R. Glennerster, D. Goroff, D. P. Green, B. Hesse, M. Humphreys, J. Ishiyama, D. Karlan, A. Kraut, A. Lupia, P. Mabry, T. Madon, N. Malhotra, E. Mayo-Wilson, M. McNutt, E. Miguel, E. L. Paluck, U. Simonsohn, C. Soderberg, B. A. Spellman, J. Turitto, G. VandenBos, S. Vazire, E. J. Wagenmakers, R. Wilson, and T. Yarkoni. 2015. Promoting an open research culture. *Science* 348(6242):1422-1425. https://doi.org/10.1126/science.aab2374.

NRC (National Research Council). 2010. *Verifying Greenhouse Gas Emissions: Methods to Support International Climate Agreements*. Washington, DC: The National Academies Press. https://doi.org/ 10.17226/12883.

Oda, T., S. Maksyutov, and R. J. Andres. 2018. The open-source data inventory for anthropogenic CO_2, version 2016 (ODIAC2016): A global monthly fossil fuel CO_2 gridded emissions data product for tracer transport simulations and surface flux inversions. *Earth System Science Data* 10(1):87-107. https://doi.org/10.5194/essd-10-87-2018.

Oda, T., R. Bun, V. Kinakh, P. Topylko, M. Halushchak, G. Marland, T. Lauvaux, M. Jonas, S. Maksyutov, Z. Nahorski, M. Lesiv, O. Danylo, and J. Horabik-Pyzel. 2019. Errors and uncertainties in a gridded carbon dioxide emissions inventory. *Mitigation and Adaptation Strategies for Global Change* 24(6):1007-1050. https://doi.org/10.1007/s11027-019-09877-2.

Oda, T., C. Haga, K. Hosomi, T. Matsui, and R. Bun. 2021. Errors and uncertainties associated with the use of unconventional activity data for estimating CO_2 emissions: The case for traffic emissions in Japan. *Environmental Research Letters* 16(8):084058. https://doi.org/10.1088/1748-9326/ac109d.

OECD (Organisation for Economic Co-operation and Development). 2015. *Climate Change Disclosure In G20 Countries: Stocktaking of corporate reporting schemes*. Paris, France: Organisation for Economic Co-operation and Development. https://www.oecd.org/investment/corporate-climate-change-disclosure-report.htm.

Oertel, C., J. Matschullat, K. Zurba, F. Zimmermann, and S. Erasmi. 2016. Greenhouse gas emissions from soils—A review. *Geochemistry* 76(3):327-352. https://doi.org/10.1016/j.chemer.2016.04.002.

Olsen, S. C., D. J. Wuebbles, and B. Owen. 2013. Comparison of global 3-D aviation emissions datasets. *Atmosperic Chemistry and Physics* 13(1):429-441. https://doi.org/10.5194/acp-13-429-2013.

Omara, M., D. Zavala-Araiza, D. R. Lyon, B. Hmiel, K. A. Roberts, and S. P. Hamburg. 2022. Methane emissions from US low production oil and natural gas well sites. *Nature Communications* 13(1):2085. https://doi.org/10.1038/s41467-022-29709-3.

O'Rourke, P., S. J. Smith, A. R. Mott, H. Ahsan, E. E. Mcduffie, M. Crippa, Z. Klimont, B. Mcdonald, S. Wang, M. B. Nicholson, R. M. Hoesly, and L. Feng. 2021. CEDS v_2021_04_21 Gridded emissions data. https://www.osti.gov/dataexplorer/biblio/dataset/1779095.

Pablo-Romero, M. d. P., R. Pozo-Barajas, and A. Sánchez-Braza. 2018. Analyzing the effects of the benchmark local initiatives of Covenant of Mayors signatories. *Journal of Cleaner Production* 176:159-174. https://doi.org/10.1016/j.jclepro.2017.12.124.

Pachauri, R. K., M. R. Allen, V. R. Barros, J. Broome, W. Cramer, R. Christ, J. A. Church, L. Clarke, Q. Dahe, P. Dasgupta, N. K. Dubash, O. Edenhofer, I. Elgizouli, C. B. Field, P. Forster, P. Friedlingstein, J. Fuglestvedt, L. Gomez-Echeverri, S. Hallegatte, G. Hegerl, M. Howden, K. Jiang, B. J. Cisneroz, V. Kattsov, H. Lee, K. J. Mach, J. Marotzke, M. D. Mastrandrea, L. Meyer, J. Minx, Y. Mulugetta, K. O'Brien, M. Oppenheimer, J. J. Pereira, R. Pichs-Madruga, G. K. Plattner, H. O. Pörtner, S. B. Power, B. Preston, N. H. Ravindranath, A. Reisinger, K. Riahi, M. Rusticucci, R. Scholes, K. Seyboth, Y. Sokona, R. Stavins, T. F. Stocker, P. Tschakert, D. v. Vuuren, and J. P. v. Ypserle. 2014. *Climate Change 2014: Synthesis Report. Contribution of Working Groups I, II and III to the Fifth Assessment Report of the Intergovernmental Panel on Climate Change*. R. K. Pachauri and L. A. Meyer, eds. Geneva, Switzerland: IPCC.

Palermo, V., P. Bertoldi, M. Apostolou, A. Kona, and S. Rivas. 2020. Assessment of climate change mitigation policies in 315 cities in the Covenant of Mayors initiative. *Sustainable Cities and Society* 60:102258. https://doi.org/10.1016/j.scs.2020.102258.

Palmer, P. I., P. Suntharalingam, D. B. A. Jones, D. J. Jacob, D. G. Streets, Q. Fu, S. A. Vay, and G. W. Sachse. 2006. Using CO_2:CO correlations to improve inverse analyses of carbon fluxes. *Journal of Geophysical Research: Atmospheres* 111(D12). https://doi.org/10.1029/2005JD006697.

Papale, D., G. Antoniella, G. Nicolini, B. Gioli, A. Zaldei, R. Vogt, C. Feigenwinter, S. Stagakis, N. Chrysoulakis, L. Järvi, E. Nemitz, C. Helfter, J. Barlow, F. Meier, E. Velasco, A. Christen, and V. Masson. 2020. Clear evidence of reduction in urban CO_2 emissions as a result of COVID-19 lockdown across Europe. https://www.researchgate.net/publication/342392861_Clear_evidence_of_reduction_in_urban_CO_2_emissions_as_a_result_of_COVID-19_lockdown_across_Europe.

Parkinson, S. 2020. The carbon boot-print of the military. *Responsible Science* (2).

Pereira, P., F. Bašić, I. Bogunovic, and D. Barcelo. 2022. Russian-Ukrainian war impacts the total environment. *Science of the Total Environment* 837:155865. https://doi.org/10.1016/j.scitotenv.2022.155865.

Pérez-Martínez, P. J., R. M. Miranda, M. F. Andrade, and P. Kumar. 2020. Air quality and fossil fuel driven transportation in the Metropolitan Area of São Paulo. *Transportation Research Interdisciplinary Perspectives* 5:100137. https://doi.org/10.1016/j.trip.2020.100137.

Perugini, L., G. Pellis, G. Grassi, P. Ciais, H. Dolman, J. I. House, G. P. Peters, P. Smith, D. Günther, and P. Peylin. 2021. Emerging reporting and verification needs under the Paris Agreement: How can the research community effectively contribute? *Environmental Science & Policy* 122:116-126. https://doi.org/10.1016/j.envsci.2021.04.012.

Peters, G. P. 2008. From production-based to consumption-based national emission inventories. *Ecological Economics* 65(1):13-23. https://doi.org/10.1016/j.ecolecon.2007.10.014.

Peters, G. P., R. M. Andrew, J. G. Canadell, P. Friedlingstein, R. B. Jackson, J. I. Korsbakken, C. Le Quéré, and A. Peregon. 2020. Carbon dioxide emissions continue to grow amidst slowly emerging climate policies. *Nature Climate Change* 10(1):3-6. https://doi.org/10.1038/s41558-019-0659-6.

Petrescu, A. M. R., G. P. Peters, G. Janssens-Maenhout, P. Ciais, F. N. Tubiello, G. Grassi, G. J. Nabuurs, A. Leip, G. Carmona-Garcia, W. Winiwarter, L. Höglund-Isaksson, D. Günther, E. Solazzo, A. Kiesow, A. Bastos, J. Pongratz, J. E. M. S. Nabel, G. Conchedda, R. Pilli, R. M. Andrew, M. J. Schelhaas, and A. J. Dolman. 2020. European anthropogenic AFOLU greenhouse gas emissions: A review and benchmark data. *Earth System Science Data* 12(2):961-1001. https://doi.org/10.5194/essd-12-961-2020.

Petrescu, A. M. R., M. J. McGrath, R. M. Andrew, P. Peylin, G. P. Peters, P. Ciais, G. Broquet, F. N. Tubiello, C. Gerbig, J. Pongratz, G. Janssens-Maenhout, G. Grassi, G. J. Nabuurs, P. Regnier, R. Lauerwald, M. Kuhnert, J. Balkovič, M. J. Schelhaas, H. A. C. Denier van der Gon, E. Solazzo, C. Qiu, R. Pilli, I. B. Konovalov, R. A. Houghton, D. Günther, L. Perugini, M. Crippa, R. Ganzenmüller, I. T. Luijkx, P. Smith, S. Munassar, R. L. Thompson, G. Conchedda, G. Monteil, M. Scholze, U. Karstens, P. Brockmann, and A. J. Dolman. 2021. The consolidated European synthesis of CO_2 emissions and removals for the European Union and United Kingdom: 1990–2018. *Earth System Science Data* 13(5):2363-2406. https://doi.org/10.5194/essd-13-2363-2021.

Peylin, P. 2022. Emerging Approaches and Integration of Multiple Data Sources. Presented at Greenhouse Gas Emissions Monitoring, Inventories, and Data Integration: Understanding the Landscape, Washington, DC, June 2, 2022.

Peylin, P., R. M. Law, K. R. Gurney, F. Chevallier, A. R. Jacobson, T. Maki, Y. Niwa, P. K. Patra, W. Peters, P. J. Rayner, C. Rödenbeck, I. T. van der Laan-Luijkx, and X. Zhang. 2013. Global atmospheric carbon budget: Results from an ensemble of atmospheric CO_2 inversions. *Biogeosciences* 10(10):6699-6720. https://doi.org/10.5194/bg-10-6699-2013.

Pfadt-Trilling, A. R., and M.-O. P. Fortier. 2021. Greenwashed energy transitions: Are US cities accounting for the life cycle greenhouse gas emissions of energy resources in climate action plans? *Energy and Climate Change* 2:100020. https://doi.org/10.1016/j.egycc.2020.100020.

Phillips, N. G., R. Ackley, E. R. Crosson, A. Down, L. R. Hutyra, M. Brondfield, J. D. Karr, K. Zhao, and R. B. Jackson. 2013. Mapping urban pipeline leaks: Methane leaks across Boston. *Environmental Pollution* 173:1-4. https://doi.org/10.1016/j.envpol.2012.11.003.

Pickers, P. A., A. C. Manning, C. Le Quéré, G. L. Forster, I. T. Luijkx, C. Gerbig, L. S. Fleming, and W. T. Sturges. 2022. Novel quantification of regional fossil fuel CO_2 reductions during COVID-19 lockdowns using atmospheric oxygen measurements. *Science Advances* 8(16):eabl9250. https://doi.org/10.1126/sciadv.abl9250.

Pinty, B., G. Janssens-Maenhout, M. Dowell, H. Zunker, T. Brunhes, P. Ciais, D. Dee, H. D. v. d. Gon, H. Dolman, M. Drinkwater, R. Engelen, M. Heimann, K. Holmlund, H. Husband, A. Kentarchos, Y. Meijer, P. Palmer, and M. Scholze. 2017. *An Operational Anthropogenic CO_2 Emissions Monitoring & Verification Support Capacity: Baseline Requirements, Model Components and Functional Architecture*. Brussels: European Commission Joint Research Centre. https://doi.org/10.2760/39384.

Pinty, B., P. Ciais, D. Dee, H. Dolman, M. Dowell, R. Engelen, K. Holmlund, G. Janssens-Maenhout, Y. Meijer, P. Palmer, M. Scholze, H. D. v. d. Gon, M. Heimann, O. Juvyns, A. Kentarchos, and H. Zunker. 2019. *An Operational Anthropogenic CO_2 Emissions Monitoring & Verification Support Capacity: Needs and High Level Requirements for In Situ Measurements*. Brussels: European Commission Joint Research Centre. https://doi.org/10.2760/182790.

Pitt, J. R., I. Lopez-Coto, K. D. Hajny, J. Tomlin, R. Kaeser, T. Jayarathne, B. H. Stirm, C. R. Floerchinger, C. P. Loughner, C. K. Gately, L. R. Hutyra, K. R. Gurney, G. S. Roest, J. Liang, S. Gourdji, A. Karion, J. R. Whetstone, and P. B. Shepson. 2022. New York City greenhouse gas emissions estimated with inverse modeling of aircraft measurements. *Elementa: Science of the Anthropocene* 10(1). https://doi.org/10.1525/elementa.2021.00082.

Plant, G., E. A. Kort, C. Floerchinger, A. Gvakharia, I. Vimont, and C. Sweeney. 2019. Large fugitive methane emissions from urban centers along the U.S. East Coast. *Geophysical Research Letters* 46(14):8500-8507. https://doi.org/10.1029/2019GL082635.

Platt, M., J. Scdlmeir, D. Platt, J. Xu, P. Tasca, N. Vadgama, and J. I. Ibañez. 2021. The energy footprint of blockchain consensus mechanisms beyond proof-of-work. Presented at IEEE 21st International Conference on Software Quality, Reliability and Security Companion (QRS-C).

Quaas, J., H. Jia, C. Smith, A. L. Albright, W. Aas, N. Bellouin, O. Boucher, M. Doutriaux-Boucher, P. M. Forster, D. Grosvenor, S. Jenkins, Z. Klimont, N. G. Loeb, X. Ma, V. Naik, F. Paulot, P. Stier, M. Wild, G. Myhre, and M. Schulz. 2022. Robust evidence for reversal in the aerosol effective climate forcing trend. *Atmospheric Chemistry and Physics Discussions* 2022:1-25. https://doi.org/10.5194/acp-2022-295.

Quick, J. C. 2014. Carbon dioxide emission tallies for 210 U.S. coal-fired power plants: A comparison of two accounting methods. *Journal of the Air and Waste Management Association* 64(1):73-79. https://doi.org/10.1080/10962247.2013.833146.

Ramachandran, A., and D. Kantarcioglu. 2017. Using blockchain and smart contracts for secure data provenance management. *arXiv preprint* arXiv:1709.10000. https://doi.org/10.48550/arXiv.1709.10000.

Ramonet, M., P. Ciais, T. Aalto, C. Aulagnier, F. Chevallier, D. Cipriano, T. J. Conway, L. Haszpra, V. Kazan, F. Meinhardt, J.-D. Paris, M. Schmidt, P. Simmonds, I. Xueref-Rémy, and J. N. Necki. 2010. A recent build-up of atmospheric CO_2 over Europe. Part 1: Observed signals and possible explanations. *Tellus B* 62(1):1-13. https://doi.org/10.1111/j.1600-0889.2009.00442.x.

Ravikumar, A. P., J. Wang, and A. R. Brandt. 2017. Are optical gas imaging technologies effective for methane leak detection? *Environmental Science & Technology* 51(1):718-724. https://doi.org/10.1021/acs.est.6b03906.

Rayner, P. J., M. Scholze, W. Knorr, T. Kaminski, R. Giering, and H. Widmann. 2005. Two decades of terrestrial carbon fluxes from a carbon cycle data assimilation system (CCDAS). *Global Biogeochemical Cycles* 19(2). https://doi.org/10.1029/2004GB002254.

Rayner, P. J., R. M. Law, C. E. Allison, R. J. Francey, C. M. Trudinger, and C. Pickett-Heaps. 2008. Interannual variability of the global carbon cycle (1992–2005) inferred by inversion of atmospheric CO_2 and $\delta^{13}CO_2$ measurements. *Global Biogeochemical Cycles* 22(3). https://doi.org/10.1029/2007GB003068.

Rebmann, C., M. Aubinet, H. Scmid, N. Arriga, M. Aurela, G. Burba, R. Clement, A. D. Ligne, G. Fratini, B. Gielen, J. Grace, A. Graf, P. Gross, S. Haapanala, M. Herbst, L. Hörtnagl, A. Ibrom, L. Joly, N. Kljun, O. K. l. Kowalski, A. Lindroth, D. Loustau, I. Mammarella, M. Mauder, L. Merbold, S. Metzger, M. Mölder, L. Montagnani, D. Papale, M. Pavelka, M. Peichl, M. Roland, P. Serrano-Ortiz, L. Siebicke, R. Steinbrecher, J. Tuovinen, T. Vesala, G. Wohlfahrt, and D. Franz. 2018. ICOS eddy covariance flux-station site setup: A review. *International Agrophysics* 32(4):471-494. https://doi.org/10.1515/intag-2017-0044.

Reckien, D., M. Salvia, O. Heidrich, J. M. Church, F. Pietrapertosa, S. De Gregorio-Hurtado, V. D'Alonzo, A. Foley, S. G. Simoes, E. Krkoška Lorencová, H. Orru, K. Orru, A. Wejs, J. Flacke, M. Olazabal, D. Geneletti, E. Feliu, S. Vasilie, C. Nador, A. Krook-Riekkola, M. Matosović, P. A. Fokaides, B. I. Ioannou, A. Flamos, N.-A. Spyridaki, M. V. Balzan, O. Fülöp, I. Paspaldzhiev, S. Grafakos, and R. Dawson. 2018. How are cities planning to respond to climate change? Assessment of local climate plans from 885 cities in the EU-28. *Journal of Cleaner Production* 191:207-219. https://doi.org/10.1016/j.jclepro.2018.03.220.

Reuter, M., M. Buchwitz, O. Schneising, S. Krautwurst, C. W. O'Dell, A. Richter, H. Bovensmann, and J. P. Burrows. 2019. Towards monitoring localized CO_2 emissions from space: Co-located regional CO_2 and NO_2 enhancements observed by the OCO-2 and S5P satellites. *Atmospheric Chemistry and Physics* 19(14):9371-9383. https://doi.org/10.5194/acp-19-9371-2019.

Rigby, M., S. Park, T. Saito, L. M. Western, A. L. Redington, X. Fang, S. Henne, A. J. Manning, R. G. Prinn, G. S. Dutton, P. J. Fraser, A. L. Ganesan, B. D. Hall, C. M. Harth, J. Kim, K. R. Kim, P. B. Krummel, T. Lee, S. Li, Q. Liang, M. F. Lunt, S. A. Montzka, J. Mühle, S. O'Doherty, M. K. Park, S. Reimann, P. K. Salameh, P. Simmonds, R. L. Tunnicliffe, R. F. Weiss, Y. Yokouchi, and D. Young. 2019. Increase in CFC-11 emissions from eastern China based on atmospheric observations. *Nature* 569(7757):546-550. https://doi.org/10.1038/s41586-019-1193-4.

Rißmann, M., J. Chen, G. Osterman, F. Dietrich, M. Makowski, X. Zhao, F. Hase, and M. Kiel. 2022. Comparison of OCO-2 target observations to MUCCnet—Is it possible to capture urban X_{CO2} gradients from space? *Atmospheric Measurement Techniques Discussions* 2022:1-27. https://doi.org/10.5194/amt-2022-71.

Rödenbeck, C., S. Houweling, M. Gloor, and M. Heimann. 2003. Time-dependent atmospheric CO_2 inversions based on interannually varying tracer transport. *Tellus B* 55(2):488-497. https://doi.org/10.1034/j.1600-0889.2003.00033.x.

REFERENCES

Rolnick, D., P. L. Donti, L. H. Kaack, K. Kochanski, A. Lacoste, K. Sankaran, A. S. Ross, N. Milojevic-Dupont, N. Jaques, A. Waldman-Brown, A. S. Luccioni, T. Maharaj, E. D. Sherwin, S. K. Mukkavilli, K. P. Kording, C. P. Gomes, A. Y. Ng, D. Hassabis, J. C. Platt, F. Creutzig, J. Chayes, and Y. Bengio. 2022. Tackling climate change with machine learning. *ACM Computing Surveys* 55(2):Article 42. https://doi.org/10.1145/3485128.

Roman-White, S. A., J. A. Littlefield, K. G. Fleury, D. T. Allen, P. Balcombe, K. E. Konschnik, J. Ewing, G. B. Ross, and F. George. 2021. LNG supply chains: a supplier-specific life-cycle assessment for improved emission accounting. *ACS Sustainable Chemistry & Engineering* 9(32):10857-10867. https://doi.org/10.1021/acssuschemeng.1c03307.

Roobaert, A., G. G. Laruelle, P. Landschützer, and P. Regnier. 2018. Uncertainty in the global oceanic CO_2 uptake induced by wind forcing: Quantification and spatial analysis. *Biogeosciences* 15(6):1701-1720. https://doi.org/10.5194/bg-15-1701-2018.

Roscioli, J. R., T. I. Yacovitch, C. Floerchinger, A. L. Mitchell, D. S. Tkacik, R. Subramanian, D. M. Martinez, T. L. Vaughn, L. Williams, D. Zimmerle, A. L. Robinson, S. C. Herndon, and A. J. Marchese. 2015. Measurements of methane emissions from natural gas gathering facilities and processing plants: Measurement methods. *Atmospheric Measurement Techniques* 8(5):2017-2035. https://doi.org/10.5194/amt-8-2017-2015.

Rosenzweig, C., W. Solecki, S. A. Hammer, and S. Mehrotra. 2010. Cities lead the way in climate–change action. *Nature* 467(7318):909-911. https://doi.org/10.1038/467909a.

Rupasinghe, R., B. B. Chomel, and B. Martínez-López. 2022. Climate change and zoonoses: A review of the current status, knowledge gaps, and future trends. *Acta Tropica* 226:106225. https://doi.org/10.1016/j.actatropica.2021.106225.

Rypdal, K., N. Paciornik, S. Eggleston, J. Goodwin, W. Irving, J. Penman, and M. Woodfield. 2006. Introduction to the 2006 Guidelines In *2006 IPCC Guidelines for National Greenhouse Gas Inventories*. S. Eggleston, L. Buendia, K. Miwa, T. Ngara, and K. Tanabe, eds. Hayama, Kanagawa, Japan: Institute for Global Environmental Strategies. https://www.ipcc-nggip.iges.or.jp/public/2006gl/pdf/1_Volume1/V1_1_Ch1_Introduction.pdf.

Saboya, E., G. Zazzeri, H. Graven, A. J. Manning, and S. Englund Michel. 2022. Continuous CH_4 and $\delta^{13}CH_4$ measurements in London demonstrate under-reported natural gas leakage. *Atmospheric Chemistry and Physics* 22(5):3595-3613. https://doi.org/10.5194/acp-22-3595-2022.

Salcedo-Sanz, S., P. Ghamisi, M. Piles, M. Werner, L. Cuadra, A. Moreno-Martínez, E. Izquierdo-Verdiguier, J. Muñoz-Marí, A. Mosavi, and G. Camps-Valls. 2020. Machine learning information fusion in Earth observation: A comprehensive review of methods, applications and data sources. *Information Fusion* 63:256-272. https://doi.org/10.1016/j.inffus.2020.07.004.

Sargent, M. R., C. Floerchinger, K. McKain, J. Budney, E. W. Gottlieb, L. R. Hutyra, J. Rudek, and S. C. Wofsy. 2021. Majority of US urban natural gas emissions unaccounted for in inventories. *Proceedings of the National Academy of Sciences* 118(44):e2105804118. https://doi.org/doi:10.1073/pnas.2105804118.

Saunois, M., A. R. Stavert, B. Poulter, P. Bousquet, J. G. Canadell, R. B. Jackson, P. A. Raymond, E. J. Dlugokencky, S. Houweling, P. K. Patra, P. Ciais, V. K. Arora, D. Bastviken, P. Bergamaschi, D. R. Blake, G. Brailsford, L. Bruhwiler, K. M. Carlson, M. Carrol, S. Castaldi, N. Chandra, C. Crevoisier, P. M. Crill, K. Covey, C. L. Curry, G. Etiope, C. Frankenberg, N. Gedney, M. I. Hegglin, L. Höglund-Isaksson, G. Hugelius, M. Ishizawa, A. Ito, G. Janssens-Maenhout, K. M. Jensen, F. Joos, T. Kleinen, P. B. Krummel, R. L. Langenfelds, G. G. Laruelle, L. Liu, T. Machida, S. Maksyutov, K. C. McDonald, J. McNorton, P. A. Miller, J. R. Melton, I. Morino, J. Müller, F. Murguia-Flores, V. Naik, Y. Niwa, S. Noce, S. O'Doherty, R. J. Parker, C. Peng, S. Peng, G. P. Peters, C. Prigent, R. Prinn, M. Ramonet, P. Regnier, W. J. Riley, J. A. Rosentreter, A. Segers, I. J. Simpson, H. Shi, S. J. Smith, L. P. Steele, B. F. Thornton, H. Tian, Y. Tohjima, F. N. Tubiello, A. Tsuruta, N. Viovy, A. Voulgarakis, T. S. Weber, M. van Weele, G. R. van der Werf, R. F. Weiss, D. Worthy, D. Wunch, Y. Yin, Y. Yoshida, W. Zhang, Z. Zhang, Y. Zhao, B. Zheng, Q. Zhu, Q. Zhu, and Q. Zhuang. 2020. The global methane budget 2000–2017. *Earth System Science Data* 12(3):1561-1623. https://doi.org/10.5194/essd-12-1561-2020.

Schaltegger, S., and M. Csutora. 2012. Carbon accounting for sustainability and management. Status quo and challenges. *Journal of Cleaner Production* 36:1-16. https://doi.org/10.1016/j.jclepro.2012.06.024.

Schelhaas, M.-J., G. M. Hengeveld, N. Heidema, E. Thürig, B. Rohner, G. Vacchiano, J. Vayreda, J. Redmond, J. Socha, J. Fridman, S. Tomter, H. Polley, S. Barreiro, and G.-J. Nabuurs. 2018. Species-specific, pan-European diameter increment models based on data of 2.3 million trees. *Forest Ecosystems* 5(1):21. https://doi.org/10.1186/s40663-018-0133-3.

Schletz, M., A. Hsu, B. Mapes, and M. Wainstein. 2022. Nested climate accounting for our atmospheric commons—digital technologies for trusted interoperability across fragmented systems. *Frontiers in Blockchain* 49. https://doi.org/10.3389/fbloc.2021.789953.

Schmidt, M., H. Glatzel-Mattheier, H. Sartorius, D. E. Worthy, and I. Levin. 2001. Western European N_2O emissions: A top-down approach based on atmospheric observations. *Journal of Geophysical Research: Atmospheres* 106(D6):5507-5516. https://doi.org/10.1029/2000JD900701.

Schmidt, M., R. Graul, H. Sartorius, and I. Levin. 2003. The Schauinsland CO_2 record: 30 years of continental observations and their implications for the variability of the European CO_2 budget. *Journal of Geophysical Research: Atmospheres* 108(D19). https://doi.org/10.1029/2002JD003085.

Schuck, T. J., C. A. M. Brenninkmeijer, F. Slemr, I. Xueref-Remy, and A. Zahn. 2009. Greenhouse gas analysis of air samples collected onboard the CARIBIC passenger aircraft. *Atmospheric Measurement Techniques* 2(2):449-464. https://doi.org/10.5194/amt-2-449-2009.

Schuh, A. E., A. R. Jacobson, S. Basu, B. Weir, D. Baker, K. Bowman, F. Chevallier, S. Crowell, K. J. Davis, F. Deng, S. Denning, L. Feng, D. Jones, J. Liu, and P. I. Palmer. 2019. Quantifying the Iimpact of atmospheric transport uncertainty on CO_2 surface flux estimates. *Global Biogeochemical Cycles* 33(4):484-500. https://doi.org/10.1029/2018GB006086.

Schwanitz, V. J., A. Wierling, M. E. Biresselioglu, M. Celino, M. H. Demir, M. Bałazińska, M. Kruczek, M. Paier, and D. Suna. 2022. Current state and call for action to accomplish findability, accessibility, interoperability, and reusability of low carbon energy data. *Scientific Reports* 12(1):5208. https://doi.org/10.1038/s41598-022-08774-0.

Schwietzke, S., O. A. Sherwood, L. M. P. Bruhwiler, J. B. Miller, G. Etiope, E. J. Dlugokencky, S. E. Michel, V. A. Arling, B. H. Vaughn, J. W. C. White, and P. P. Tans. 2016. Upward revision of global fossil fuel methane emissions based on isotope database. *Nature* 538(7623):88-91. https://doi.org/10.1038/nature19797.

Schwietzke, S., G. Pétron, S. Conley, C. Pickering, I. Mielke-Maday, E. J. Dlugokencky, P. P. Tans, T. Vaughn, C. Bell, D. Zimmerle, S. Wolter, C. W. King, A. B. White, T. Coleman, L. Bianco, and R. C. Schnell. 2017. Improved mechanistic understanding of natural gas methane emissions from spatially resolved aircraft measurements. *Environmental Science & Technology* 51(12):7286-7294. https://doi.org/10.1021/acs.est.7b01810.

SEC (Securities and Exchange Commission). 2022. The Enhancement and Standardization of Climate-Related Disclosures for Investors A Proposed Rule by the Securities and Exchange Commission on 05/12/2022. Document number 2022-10194. *Federal Register* 87 FR 29059. https://www.federalregister.gov/d/22022-10194.

Sedlmeir, J., H. U. Buhl, G. Fridgen, and R. Keller. 2020. The energy consumption of blockchain technology: Beyond myth. *Business & Information Systems Engineering* 62(6):599-608. https://doi.org/10.1007/s12599-020-00656-x.

Seto, K. C., S. Dhakal, A. Bigio, H. Blanco, G. C. Delgado, D. Dewar, L. Huang, A. Inaba, A. Kansal, S. Lwasa, J. McMahon, D. B. Müller, J. Murakami, H. Nagendra, and A. Ramaswami. 2014. Human settlements, infrastructure and spatial planning. In *Climate Change 2014: Mitigation of Climate Change. Contribution of Working Group III to the Fifth Assessment Report of the Intergovernmental Panel on Climate Change*. O. Edenhofer, R. Pichs-Madruga, Y. Sokona, E. Farahani, S. Kadner, K. Seyboth, A. Adler, I. Baum, S. Brunner, P. Eickemeier, B. Kriemann, J. Savolainen, S. Schlömer, C. v. Stechow, T. Zwickel, and J. C. Minx, eds. Cambridge, UK and New York, NY: Cambridge University Press.

Sha, Z., Y. Bai, R. Li, H. Lan, X. Zhang, J. Li, X. Liu, S. Chang, and Y. Xie. 2022. The global carbon sink potential of terrestrial vegetation can be increased substantially by optimal land management. *Communications Earth & Environment* 3(1):8. https://doi.org/10.1038/s43247-021-00333-1.

Shaddick, G., M. L. Thomas, A. Green, M. Brauer, A. van Donkelaar, R. Burnett, H. H. Chang, A. Cohen, R. V. Dingenen, C. Dora, S. Gumy, Y. Liu, R. Martin, L. A. Waller, J. West, J. V. Zidek, and A. Prüss-Ustün. 2018. Data integration model for air quality: A hierarchical approach to the global estimation of exposures to ambient air pollution. *Journal of the Royal Statistical Society: Series C (Applied Statistics)* 67(1):231-253. https://doi.org/10.1111/rssc.12227.

Shan, Y., D. Guan, H. Zheng, J. Ou, Y. Li, J. Meng, Z. Mi, Z. Liu, and Q. Zhang. 2018. China CO_2 emission accounts 1997–2015. *Scientific Data* 5(1):170201. https://doi.org/10.1038/sdata.2017.201.

Shan, Y., Q. Huang, D. Guan, and K. Hubacek. 2020. China CO_2 emission accounts 2016–2017. *Scientific Data* 7(1):54. https://doi.org/10.1038/s41597-020-0393-y.

Shen, L., R. Gautam, M. Omara, D. Zavala-Araiza, J. Maasakkers, T. Scarpelli, A. Lorente, D. Lyon, J. Sheng, D. Varon, H. Nesser, Z. Qu, X. Lu, M. Sulprizio, S. Hamburg, and D. Jacob. 2022. Satellite quantification of oil and natural gas methane emissions in the US and Canada including contributions from individual basins. *Atmospheric Chemistry and Physics*, 22, 11203–11215, https://doi.org/10.5194/acp-22-11203-2022.

Shusterman, A. A., V. E. Teige, A. J. Turner, C. Newman, J. Kim, and R. C. Cohen. 2016. The BErkeley Atmospheric CO_2 Observation Network: Initial evaluation. *Atmospheric Chemistry and Physics* 16(21):13449-13463. https://doi.org/10.5194/acp-16-13449-2016.

Siciliano, B., G. Dantas, C. M. da Silva, and G. Arbilla. 2020. Increased ozone levels during the COVID-19 lockdown: Analysis for the city of Rio de Janeiro, Brazil. *Science of the Total Environment* 737:139765. https://doi.org/10.1016/j.scitotenv.2020.139765.

Smith, N. E., L. M. J. Kooijmans, G. Koren, E. van Schaik, A. M. van der Woude, N. Wanders, M. Ramonet, I. Xueref-Remy, L. Siebicke, G. Manca, C. Brümmer, I. T. Baker, K. D. Haynes, I. T. Luijkx, and W. Peters. 2020. Spring enhancement and summer reduction in carbon uptake during the 2018 drought in northwestern Europe. *Philosophical Transactions of the Royal Society B: Biological Sciences* 375(1810):20190509. https://doi.org/10.1098/rstb.2019.0509.

Solazzo, E., M. Crippa, D. Guizzardi, M. Muntean, M. Choulga, and G. Janssens-Maenhout. 2021. Uncertainties in the Emissions Database for Global Atmospheric Research (EDGAR) emission inventory of greenhouse gases. *Atmospheric Chemistry and Physics* 21(7):5655-5683. https://doi.org/10.5194/acp-21-5655-2021.

Sporny, M., D. Longley, and D. Chadwick. 2022. Verifiable Credentials Data Model 1.1. W3C Recommendation 03 March 2022. https://www.w3.org/TR/vc-data-model.

Stanley, K. M., D. Say, J. Mühle, C. M. Harth, P. B. Krummel, D. Young, S. J. O'Doherty, P. K. Salameh, P. G. Simmonds, R. F. Weiss, R. G. Prinn, P. J. Fraser, and M. Rigby. 2020. Increase in global emissions of HFC-23 despite near-total expected reductions. *Nature Communications* 11(1):397. https://doi.org/10.1038/s41467-019-13899-4.

Stern, J. 2022. *Measurement, Reporting, and Verification of Methane Emissions from Natural Gas and LNG Trade: Creating Transparent and Credible Frameworks.* Oxford, UK: Oxford Institute for Energy Studies. https://www.oxfordenergy.org/wpcms/wp-content/uploads/2022/01/Measurement-Reporting-and-Verification-of-Methane-Emissions-from-Natural-Gas-and-LNG-Trade-ET06.pdf.

Subramanian, R., L. L. Williams, T. L. Vaughn, D. Zimmerle, J. R. Roscioli, S. C. Herndon, T. I. Yacovitch, C. Floerchinger, D. S. Tkacik, A. L. Mitchell, M. R. Sullivan, T. R. Dallmann, and A. L. Robinson. 2015. Methane emissions from natural gas compressor stations in the transmission and storage sector: Measurements and comparisons with the EPA greenhouse gas reporting program protocol. *Environmental Science & Technology* 49(5):3252-3261. https://doi.org/10.1021/es5060258.

Sudmanns, M., D. Tiede, S. Lang, H. Bergstedt, G. Trost, H. Augustin, A. Baraldi, and T. Blaschke. 2020. Big Earth data: Disruptive changes in Earth observation data management and analysis? *International Journal of Digital Earth* 13(7):832-850. https://doi.org/10.1080/17538947.2019.1585976.

Super, I., H. A. C. Denier van der Gon, M. K. van der Molen, S. N. C. Dellaert, and W. Peters. 2020. Optimizing a dynamic fossil fuel CO_2 emission model with CTDAS (CarbonTracker Data Assimilation Shell, v1.0) for an urban area using atmospheric observations of CO_2, CO, NO_x, and SO_2. *Geoscientific Model Development* 13(6):2695-2721. https://doi.org/10.5194/gmd-13-2695-2020.

Susiluoto, J., M. Raivonen, L. Backman, M. Laine, J. Makela, O. Peltola, T. Vesala, and T. Aalto. 2018. Calibrating the sqHIMMELI v1.0 wetland methane emission model with hierarchical modeling and adaptive MCMC. *Geoscientific Model Development* 11(3):1199-1228. https://doi.org/10.5194/gmd-11-1199-2018.

Szopa, S., V. Naik, B. Adhikary, P. Artaxo, T. Berntsen, W. D. Collins, S. Fuzzi, L. Gallardo, A. Kiendler-Scharr, Z. Klimont, H. Liao, N. Unger, and P. Zanis. 2021. Short-lived climate forcers. In *Climate Change 2021: The Physical Science Basis. Contribution of Working Group I to the Sixth Assessment Report of the Intergovernmental Panel on Climate Change.* V. Masson-Delmotte, P. Zhai, A. Pirani, S. L. Connors, C. Péan, S. Berger, N. Caud, Y. Chen, L. Goldfarb, M. I. Gomis, M. Huang, K. Leitzell, E. Lonnoy, J. B. R. Matthews, T. K. Maycock, T. Waterfield, O. Yelekçi, R. Yu, and B. Zhou, eds. Cambridge, United Kingdom and New York, NY, USA: Cambridge University Press.

Tans, P. P., I. Y. Fung, and T. Takahashi. 1990. Observational contrains on the global atmospheric CO_2 budget. *Science* 247(4949):1431-1438. https://doi.org/10.1126/science.247.4949.1431.

Tarantola, A. 1987. *Inverse Problem Theory: Methods for Data Fitting and Model Parameter Estimation.* Amsterdam: Elsevier Science.

Thompson, R. L., L. Lassaletta, P. K. Patra, C. Wilson, K. C. Wells, A. Gressent, E. N. Koffi, M. P. Chipperfield, W. Winiwarter, E. A. Davidson, H. Tian, and J. G. Canadell. 2019. Acceleration of global N_2O emissions seen from two decades of atmospheric inversion. *Nature Climate Change* 9(12):993-998. https://doi.org/10.1038/s41558-019-0613-7.

Tollefson, J. 2022. Climate pledges from top companies crumble under scrutiny. *Nature News.* https://doi.org/10.1038/d41586-022-00366-2.

Truby, J. 2018. Decarbonizing Bitcoin: Law and policy choices for reducing the energy consumption of Blockchain technologies and digital currencies. *Energy Research & Social Science* 44:399-410. https://doi.org/10.1016/j.erss.2018.06.009.

Tsai, C. W., C. F. Lai, M. C. Chiang, and L. T. Yang. 2014. Data mining for Internet of Things: A survey. *IEEE Communications Surveys & Tutorials* 16(1):77-97. https://doi.org/10.1109/SURV.2013.103013.00206.

Tsai, T. R., K. Du, and B. Stavropoulos. 2017. New system for detecting, mapping, monitoring, quantifying and reporting fugitive gas emissions. *The APPEA Journal* 57(2):561-566. https://doi.org/10.1071/AJ16098.

Tubiello, F. N., M. Salvatore, S. Rossi, A. Ferrara, N. Fitton, and P. Smith. 2013. The FAOSTAT database of greenhouse gas emissions from agriculture. *Environmental Research Letters* 8(1):015009. https://doi.org/10.1088/1748-9326/8/1/01500

Turnbull, J. C., E. D. Keller, T. Baisden, G. Brailsford, T. Bromley, M. Norris, and A. Zondervan. 2014. Atmospheric measurement of point source fossil CO_2 emissions. *Atmospheric Chemistry and Physics* 14(10):5001-5014. https://doi.org/10.5194/acp-14-5001-2014.

Turnbull, J. C., C. Sweeney, A. Karion, T. Newberger, S. J. Lehman, P. P. Tans, K. J. Davis, T. Lauvaux, N. L. Miles, S. J. Richardson, M. O. Cambaliza, P. B. Shepson, K. Gurney, R. Patarasuk, and I. Razlivanov. 2015. Toward quantification and source sector identification of fossil fuel CO_2 emissions from an urban area: Results from the INFLUX experiment. *Journal of Geophysical Research: Atmospheres* 120(1):292-312. https://doi.org/10.1002/2014JD022555.

Turnbull, J. C., A. Karion, K. J. Davis, T. Lauvaux, N. L. Miles, S. J. Richardson, C. Sweeney, K. McKain, S. J. Lehman, K. R. Gurney, R. Patarasuk, J. Liang, P. B. Shepson, A. Heimburger, R. Harvey, and J. Whetstone. 2019. Synthesis of urban CO_2 emission estimates from multiple methods from the Indianapolis Flux Project (INFLUX). *Environmental Science & Technology* 53(1):287-295. https://doi.org/10.1021/acs.est.8b05552.

Turnbull, J. C., L. G. Domingues, and N. Turton. 2022. Dramatic lockdown fossil fuel CO_2 decrease detected by citizen science-supported atmospheric radiocarbon observations. *Environmental Science & Technology* 56(14):9882-9890. https://doi.org/10.1021/acs.est.1c07994.

Turner, A. J., J. Kim, H. Fitzmaurice, C. Newman, K. Worthington, K. Chan, P. J. Wooldridge, P. Köehler, C. Frankenberg, and R. C. Cohen. 2020. Observed impacts of COVID-19 on urban CO_2 emissions. *Geophysical Research Letters* 47(22):e2020GL090037. https://doi.org/https://doi.org/10.1029/2020GL090037.

Umemiya, C., M. White, A. Amellina, and N. Shimizu. 2017. National greenhouse gas inventory capacity: An assessment of Asian developing countries. *Environmental Science & Policy* 78:66-73. https://doi.org/10.1016/j.envsci.2017.09.008.

UNEP (United Nations Environment Programme). 2019. *Volume 1: Decision XXXI/3 TEAP Task Force Report on Unexpected Emissions of Trichlorofluoromethane (CFC-11). Report of the Technology and Economic Assessment Panel of the Montreal Protocol on Substances that Deplete the Ozone Layer.* Nairobi, Kenya: United Nations Environment Programme. https://ozone.unep.org/system/files/documents/TEAP-TF-DecXXX-3-unexpected_CFC11_emissions-september2019.pdf.

UNEP. 2020. *From Disclosure to Action: Applying TCFD Principles throughout Financial Institutions.* Geneva, Switzerland: UNEP. https://www.unepfi.org/wordpress/wp-content/uploads/2020/10/Climate-Risk-Applications-From-Disclosure-to-Action.pdf.

UNEP. 2021a. *An Eye on Methane: International Methane Emissions Observatory 2021 Report.* Nairobi, Kenya: UNEP. https://www.unep.org/resources/report/eye-methane-international-methane-emissions-observatory-2021-report.

UNEP. 2021b. *Volume 3: Decision XXXI/3 TEAP Task Force Report on Unexpected Emissions of Trichlorofluoromethane (CFC-11). Report of the Technology and Economic Assessment Panel of the Montreal Protocol on Substances that Deplete the Ozone Layer.* Nairobi, Kenya: United Nations Environment Programme. https://ozone.unep.org/system/files/documents/Final_TEAP-DecisionXXXI-3-TF-Unexpected-Emissions-of-CFC-11-may2021.pdf.

UNEP and CCAC (Climate and Clean Air Coalition). 2021. *Global Methane Assessment: Benefits and Costs of Mitigating Methane Estimates.* Nairobi, Kenya: United Nations Environment Programme.

UNFCCC (United Nations Framework Convention on Climate Change). 1992. Status of Ratification of the Convention. https://unfccc.int/process-and-meetings/the-convention/status-of-ratification/status-of-ratification-of-the-convention.

UNFCCC. 2014. Report of the Conference of the Parties on its nineteenth session, held in Warsaw from 11 to 23 November 2013. Addendum. Part Two: Action taken by the Conference of the Parties at its nineteenth session. FCCC/CP/2013/10/Add.3. https://unfccc.int/resource/docs/2013/cop19/eng/10a03.pdf.

UNFCCC. 2015. The Paris Agreement. https://unfccc.int/sites/default/files/english_paris_agreement.pdf.

UNFCCC. 2019. Report of the Conference of the Parties serving as the meeting of the Parties to the Paris Agreement on the third part of its first session, held in Katowice from 2 to 15 December 2018. Addendum. Part Two: Action taken by the Conference of the Parties serving as the meeting of the Parties to the Paris Agreement. FCCC/PA/CMA/2018/3/Add.2. https://unfccc.int/sites/default/files/resource/CMA2018_03a02E.pdf.

UNFCCC. 2020. Preparing for Implementation of the Enhanced Transparency Framework under the Paris Agreement. https://unfccc.int/sites/default/files/resource/ETF%20Technical%20Handbook%20First%20Edition%20June_2020.pdf.

UNFCCC. 2021. National Inventory Submissions 2021. https://unfccc.int/ghg-inventories-annex-i-parties/2021.

UNFCCC. 2021b. Moving Towards the Enhanced Transparency Framework. https://unfccc.int/enhanced-transparency-framework.

UNFCCC. 2022a. National Inventory Submissions 2022. https://unfccc.int/ghg-inventories-annex-i-parties/2022.

UNFCCC. 2022b. The Role of Systematic Earth Observations in the Global Stocktake. https://www4.unfccc.int/sites/SubmissionsStaging/Documents/202203012343---SO-in-GST-2022-final.pdf.

UNFCCC. n.d.-a. Global Stocktake. https://unfccc.int/topics/global-stocktake/global-stocktake#eq-1.

UNFCCC. n.d.-b. National Communication Submissions from Non-Annex I Parties. https://unfccc.int/non-annex-I-NCs.

UNFCCC. n.d.-c. Biennial Update Report Submissions from Non-Annex I Parties. https://unfccc.int/BURs.

UNSD (United Nations Statistics Division). 2022. Background Document to the Report of the Secretary-General on Climate Change Statistics (E/CN.3/2022/17). Global Consultation on the Global Set. https://unstats.un.org/unsd/statcom/53rd-session/documents/BG-3m-GlobalConsultationontheGlobalSet-E.pdf.

van Genderen, J., M. F. Goodchild, H. Guo, C. Yang, S. Nativi, L. Wang, and C. Wang. 2020. Digital Earth challenges and future trends. In *Manual of Digital Earth.* H. Guo, M. F. Goodchild, and A. Annoni, eds. Singapore: Springer Singapore.

Varon, D. J., D. J. Jacob, J. McKeever, D. Jervis, B. O. A. Durak, Y. Xia, and Y. Huang. 2018. Quantifying methane point sources from fine-scale satellite observations of atmospheric methane plumes. *Atmospheric Measurement Techniques* 11(10):5673-5686. https://doi.org/10.5194/amt-11-5673-2018.

Varon, D. J., J. McKeever, D. Jervis, J. D. Maasakkers, S. Pandey, S. Houweling, I. Aben, T. Scarpelli, and D. J. Jacob. 2019. Satellite discovery of anomalously large methane point sources from oil/gas production. *Geophysical Research Letters* 46(22):13507-13516. https://doi.org/10.1029/2019GL083798.

Vaughn, T. L., C. S. Bell, C. K. Pickering, S. Schwietzke, G. A. Heath, G. Pétron, D. J. Zimmerle, R. C. Schnell, and D. Nummedal. 2018. Temporal variability largely explains top-down/bottom-up difference in methane emission estimates from a natural gas production region. *Proceedings of the National Academy of Sciences* 115(46):11712-11717. https://doi.org/10.1073/pnas.1805687115.

Vechi, N. T., J. Mellqvist, and C. Scheutz. 2022. Quantification of methane emissions from cattle farms, using the tracer gas dispersion method. *Agriculture, Ecosystems & Environment* 330:107885. https://doi.org/10.1016/j.agee.2022.107885.

Velasco, E., M. Roth, S. H. Tan, M. Quak, S. D. A. Nabarro, and L. Norford. 2013. The role of vegetation in the CO_2 flux from a tropical urban neighbourhood. *Atmospheric Chemistry and Physics* 13(20):10185-10202. https://doi.org/10.5194/acp-13-10185-2013.

Velasco, E., M. Roth, L. Norford, and L. T. Molina. 2016. Does urban vegetation enhance carbon sequestration? *Landscape and Urban Planning* 148:99-107. https://doi.org/10.1016/j.landurbplan.2015.12.003.

Verhulst, K. R., A. Karion, J. Kim, P. K. Salameh, R. F. Keeling, S. Newman, J. Miller, C. Sloop, T. Pongetti, P. Rao, C. Wong, F. M. Hopkins, V. Yadav, R. F. Weiss, R. M. Duren, and C. E. Miller. 2017. Carbon dioxide and methane measurements from the Los Angeles Megacity Carbon Project – Part 1: Calibration, urban enhancements, and uncertainty estimates. *Atmospheric Chemistry and Physics* 17(13):8313-8341. https://doi.org/10.5194/acp-17-8313-2017.

Vollmer, M. K., J. Mühle, S. Henne, D. Young, M. Rigby, B. Mitrevski, S. Park, C. R. Lunder, T. S. Rhee, C. M. Harth, M. Hill, R. L. Langenfelds, M. Guillevic, P. M. Schlauri, O. Hermansen, J. Arduini, R. H. J. Wang, P. K. Salameh, M. Maione, P. B. Krummel, S. Reimann, S. O'Doherty, P. G. Simmonds, P. J. Fraser, R. G. Prinn, R. F. Weiss, and L. P. Steele. 2021. Unexpected nascent atmospheric emissions of three ozone-depleting hydrochlorofluorocarbons. *Proceedings of the National Academy of Sciences* 118(5):e2010914118. https://doi.org/10.1073/pnas.2010914118.

von Fischer, J. C., D. Cooley, S. Chamberlain, A. Gaylord, C. J. Griebenow, S. P. Hamburg, J. Salo, R. Schumacher, D. Theobald, and J. Ham. 2017. Rapid, vehicle-based identification of location and magnitude of urban natural gas pipeline leaks. *Environmental Science & Technology* 51(7):4091-4099. https://doi.org/10.1021/acs.est.6b06095.

Wainstein, M. E. 2019. Open Climate. Leveraging blockchain for a global, transparent and integrated climate accounting system. Presented at Yale Open Innovation Lab: Openlab, New Haven, Connecticut. https://collabathon-docs.openclimate.earth/openclimate/docs-open-climate-platform.

Wang, H., C. Ma, and L. Zhou. 2009. A brief review of machine learning and its application. Presented at 2009 International Conference on Information Engineering and Computer Science, December 19-20.

Wang, F., S. Maksyutov, A. Tsuruta, R. Janardanan, A. Ito, M. Sasakawa, T. Machida, I. Morino, Y. Yoshida, J. W. Kaiser, G. Janssens-Maenhout, E. J. Dlugokencky, I. Mammarella, J. V. Lavric, and T. Matsunaga. 2019. Methane emission estimates by the Global High-Resolution Inverse Model using national inventories. *Remote Sensing* 11(21):2489.

Wang, Z., L. Lin, Y. Xu, H. Che, X. Zhang, H. Zhang, W. Dong, C. Wang, K. Gui, and B. Xie. 2021. Incorrect Asian aerosols affecting the attribution and projection of regional climate change in CMIP6 models. *npj Climate and Atmospheric Science* 4(1):2. https://doi.org/10.1038/s41612-020-00159-2.

Wang, J., W. Daniels, D. Hammerling, M. Harrison, K. Burmaster, F. George, and A. Ravikumar. 2022a. Multi-scale methane measurements at oil and gas facilities reveal necessary conditions for improved emissions accounting. *ChemRxiv*. https://doi.org/10.26434/chemrxiv-2022-9zh2v.

Wang, J., L. Feng, P. I. Palmer, Y. Liu, S. Fang, H. Bösch, C. W. O'Dell, X. Tang, D. Yang, L. Liu, and C. Xia. 2022b. Reply to: On the role of atmospheric model transport uncertainty in estimating the Chinese land carbon sink. *Nature* 603(7901):E15-E16. https://doi.org/10.1038/s41586-021-04259-8.

Watts, M. 2017. Cities spearhead climate action. *Nature Climate Change* 7(8):537-538. https://doi.org/10.1038/nclimate3358.

Wei, T., J. Wu, and S. Chen. 2021. Keeping track of greenhouse gas emission reduction progress and targets in 167 cities worldwide. *Frontiers in Sustainable Cities* 3. https://doi.org/10.3389/frsc.2021.696381.

Weikmans, R., and A. Gupta. 2021. Assessing state compliance with multilateral climate transparency requirements: 'Transparency Adherence Indices' and their research and policy implications. *Climate Policy* 21(5):635-651. https://doi.org/10.1080/14693062.2021.1895705.

Weir, B., D. Crisp, C. W. O'Dell, S. Basu, A. Chatterjee, J. Kolassa, T. Oda, S. Pawson, B. Poulter, Z. Zhang, P. Ciais, S. J. Davis, Z. Liu, and L. E. Ott. 2021. Regional impacts of COVID-19 on carbon dioxide detected worldwide from space. *Science Advances* 7(45):eabf9415. https://doi.org/10.1126/sciadv.abf9415.

Weissert, L. F., J. A. Salmond, and L. Schwendenmann. 2017. Photosynthetic CO_2 uptake and carbon sequestration potential of deciduous and evergreen tree species in an urban environment. *Urban Ecosystems* 20(3):663-674. https://doi.org/10.1007/s11252-016-0627-0.

Whetstone, J. R. 2018. Advances in urban greenhouse gas flux quantification: The Indianapolis Flux Experiment (INFLUX). *Elementa: Science of the Anthropocene* 6. https://doi.org/10.1525/elementa.282.

Wilkinson, M. D., M. Dumontier, I. J. Aalbersberg, G. Appleton, M. Axton, A. Baak, N. Blomberg, J.-W. Boiten, L. B. da Silva Santos, P. E. Bourne, J. Bouwman, A. J. Brookes, T. Clark, M. Crosas, I. Dillo, O. Dumon, S. Edmunds, C. T. Evelo, R. Finkers, A. Gonzalez-Beltran, A. J. G. Gray, P. Groth, C. Goble, J. S. Grethe, J. Heringa, P. A. C. 't Hoen, R. Hooft, T. Kuhn, R. Kok, J. Kok, S. J. Lusher, M. E. Martone, A. Mons, A. L. Packer, B. Persson, P. Rocca-Serra, M. Roos, R. van Schaik, S.-A. Sansone, E. Schultes, T. Sengstag, T. Slater, G. Strawn, M. A. Swertz, M. Thompson, J. van der Lei, E. van Mulligen, J. Velterop, A. Waagmeester, P. Wittenburg, K. Wolstencroft, J. Zhao, and B. Mons. 2016. The FAIR Guiding Principles for scientific data management and stewardship. *Scientific Data* 3(1):160018. https://doi.org/10.1038/sdata.2016.18.

Wilson, D., and J. Swisher. 1993. Exploring the gap: Top-down versus bottom-up analyses of the cost of mitigating global warming. *Energy Policy* 21(3):249-263. https://doi.org/10.1016/0301-4215(93)90247-D.

WMO (World Meteorological Organization). 2021. *Report on Unexpected Emissions of CFC-11: A Report of the Scientific Assessment Panel of the Montreal Protocol on Substances that Deplete the Ozone Layer.* Geneva, Switzerland: WMO. https://ozone.unep.org/system/files/documents/SAP-2021-report-on-the-unexpected-emissions-of-CFC-11-1268_en.pdf.

WMO/IAEA (International Atomic Energy Agency). 2013. 17th WMO/IAEA Meeting on Carbon Dioxide, Other Greenhouse Gases and Related Tracers Measurement Techniques (GGMT-2013). P. Tans and Christoph Zellweger, eds. https://www.uncclearn.org/wp-content/uploads/library/gaw_213_en.pdf.

WMO/IAEA. 2020. 20th WMO/IAEA Meeting on Carbon Dioxide, Other Greenhouse Gases and Related Tracers Measurement Techniques (GGMT-2019). A. Crotwell, H. Lee and M. Steinbacher, eds. https://library.wmo.int/doc_num.php?explnum_id=10353.

Worden, J. R., D. H. Cusworth, Z. Qu, Y. Yin, Y. Zhang, A. A. Bloom, S. Ma, B. K. Byrne, T. Scarpelli, J. D. Maasakkers, D. Crisp, R. Duren, and D. J. Jacob. 2022. The 2019 methane budget and uncertainties at 1° resolution and each country through Bayesian integration of GOSAT total column methane data and a priori inventory estimates. *Atmospheric Chemistry and Physics* 22(10):6811-6841. https://doi.org/10.5194/acp-22-6811-2022.

WRI (World Resources Institute). 2015. *Guide for Designing Mandatory Greenhouse Gas Reporting Programs.* Washington, DC: World Resources Institute. https://www.wri.org/research/guide-designing-mandatory-greenhouse-gas-reporting-programs.

WRI/WBCSD (World Business Council for Sustainable Development). 2004. *The Greenhouse Gas Protocol: A Corporate Accounting and Reporting Standard.* Washington, DC: World Resource Institute. https://ghgprotocol.org/sites/default/files/ standards/ghg-protocol-revised.pdf.

WRI/WBCSD. 2011. *Corporate Value Chain (Scope 3) Accounting and Reporting Standard.* Supplement to the GHG Protocol Corporate Accounting and Reporting Standard. Geneva, Switzerland: Greenhouse Gas Protocol.

Wu, K., K. J. Davis, N. L. Miles, S. J. Richardson, T. Lauvaux, D. P. Sarmiento, N. V. Balashov, K. Keller, J. Turnbull, K. R. Gurney, J. Liang, and G. Roest. 2022. Source decomposition of eddy-covariance CO_2 flux measurements for evaluating a high-resolution urban CO2 emissions inventory. *Environmental Research Letters* 17(7):074035. https://doi.org/10.1088/1748-9326/ac7c29.

Wuebbles, D., M. Gupta, and M. Ko. 2007. Evaluating the impacts of aviation on climate change. *Eos, Transactions American Geophysical Union* 88(14):157-160. https://doi.org/10.1029/2007EO140001.

Wunch, D., P. O. Wennberg, G. C. Toon, G. Keppel-Aleks, and Y. G. Yavin. 2009. Emissions of greenhouse gases from a North American megacity. *Geophysical Research Letters* 36(15). https://doi.org/10.1029/2009GL039825.

Xueref-Remy, I., P. Bousquet, C. Carouge, L. Rivier, and P. Ciais. 2011a. Variability and budget of CO_2 in Europe: Analysis of the CAATER airborne campaigns – Part 2: Comparison of CO_2 vertical variability and fluxes between observations and a modeling framework. *Atmospheric Chemistry and Physics* 11(12):5673-5684. https://doi.org/10.5194/acp-11-5673-2011.

Xueref-Remy, I., C. Messager, D. Filippi, M. Pastel, P. Nedelec, M. Ramonet, J. D. Paris, and P. Ciais. 2011b. Variability and budget of CO_2 in Europe: Analysis of the CAATER airborne campaigns – Part 1: Observed variability. *Atmospheric Chemistry and Physics* 11(12):5655-5672. https://doi.org/10.5194/acp-11-5655-2011.

Xueref-Remy, I., E. Dieudonné, C. Vuillemin, M. Lopez, C. Lac, M. Schmidt, M. Delmotte, F. Chevallier, F. Ravetta, O. Perrussel, P. Ciais, F. M. Bréon, G. Broquet, M. Ramonet, T. G. Spain, and C. Ampe. 2018. Diurnal, synoptic and seasonal variability of atmospheric CO_2 in the Paris megacity area. *Atmospheric Chemistry and Physics* 18(5):3335-3362. https://doi.org/10.5194/acp-18-3335-2018.

Xueref-Remy, I., G. Zazzeri, F. M. Bréon, F. Vogel, P. Ciais, D. Lowry, and E. G. Nisbet. 2020a. Anthropogenic methane plume detection from point sources in the Paris megacity area and characterization of their $\delta^{13}C$ signature. *Atmospheric Environment* 222:117055. https://doi.org/10.1016/j.atmosenv.2019.117055.

Xueref-Remy, I., A. Riandet, L. Lelandais, B. Nathan, M. Milne, V. Masson, M.-L. Lambert, A. Armengaud, J. Turnbull, C. Yohia, A. Nicault, T. Lauvaux, J. Piazzola, C. Lac, T. Hedde, S. Robert, G. Simioni, W. Cramer, and A. Bondeau. 2020b. COoL-AMmetropolis: Towards establishing virtuous greenhouse gas emission mitigation scenarios for 2035 in the Aix-Marseille metropolis area (France) through atmospheric top-down technics and social sciences methods in interaction with local stakeholders. Presented at EGU General Assembly 2020. https://doi.org/10.5194/egusphere-egu2020-5930.

Yacovitch, T. I., S. C. Herndon, G. Pétron, J. Kofler, D. Lyon, M. S. Zahniser, and C. E. Kolb. 2015. Mobile laboratory observations of methane emissions in the Barnett Shale Region. *Environmental Science & Technology* 49(13):7889-7895. https://doi.org/10.1021/es506352j.

Yona, L., B. Cashore, R. B. Jackson, J. Ometto, and M. A. Bradford. 2020. Refining national greenhouse gas inventories. *Ambio* 49(10):1581-1586. https://doi.org/10.1007/s13280-019-01312-9.

Yona, L., B. Cashore, and M. A. Bradford. 2022. Factors influencing the development and implementation of national greenhouse gas inventory methodologies. *Policy Design and Practice* 5(2):197-225. https://doi.org/10.1080/25741292.2021.2020967.

Yu, K., C. A. Keller, D. J. Jacob, A. M. Molod, S. D. Eastham, and M. S. Long. 2018. Errors and improvements in the use of archived meteorological data for chemical transport modeling: An analysis using GEOS-Chem v11-01 driven by GEOS-5 meteorology. *Geoscientific Model Development* 11(1):305-319. https://doi.org/10.5194/gmd-11-305-2018.

Yver-Kwok, C., C. Philippon, P. Bergamaschi, T. Biermann, F. Calzolari, H. Chen, S. Conil, P. Cristofanelli, M. Delmotte, J. Hatakka, M. Heliasz, O. Hermansen, K. Komínková, D. Kubistin, N. Kumps, O. Laurent, T. Laurila, I. Lehner, J. Levula, M. Lindauer, M. Lopez, I. Mammarella, G. Manca, P. Marklund, J. M. Metzger, M. Mölder, S. M. Platt, M. Ramonet, L. Rivier, B. Scheeren, M. K. Sha, P. Smith, M. Steinbacher, G. Vítková, and S. Wyss. 2021. Evaluation and optimization of ICOS atmosphere station data as part of the labeling process. *Atmospheric Measurement Techniques* 14(1):89-116. https://doi.org/10.5194/amt-14-89-2021.

Zavala-Araiza, D., S. C. Herndon, J. R. Roscioli, T. I. Yacovitch, M. R. Johnson, D. R. Tyner, M. Omara, and B. Knighton. 2018. Methane emissions from oil and gas production sites in Alberta, Canada. *Elementa: Science of the Anthropocene* 6. https://doi.org/10.1525/elementa.284.

Zazzeri, G., D. Lowry, R. E. Fisher, J. L. France, M. Lanoisellé, and E. G. Nisbet. 2015. Plume mapping and isotopic characterisation of anthropogenic methane sources. *Atmospheric Environment* 110:151-162. https://doi.org/10.1016/j.atmosenv.2015.03.029.

Zazzeri, G., X. Xu, and H. Graven. 2021. Efficient sampling of atmospheric methane for radiocarbon analysis and quantification of fossil methane. *Environmental Science & Technology* 55(13):8535-8541. https://doi.org/10.1021/acs.est.0c03300.

Zhang, Y., R. Gautam, S. Pandey, M. Omara, J. D. Maasakkers, P. Sadavarte, D. Lyon, H. Nesser, M. P. Sulprizio, D. J. Varon, R. Zhang, S. Houweling, D. Zavala-Araiza, R. A. Alvarez, A. Lorente, S. P. Hamburg, I. Aben, and D. J. Jacob. 2020. Quantifying methane emissions from the largest oil-producing basin in the United States from space. *Science Advances* 6(17):eaaz5120. https://doi.org/10.1126/sciadv.aaz5120.

Zhao, Y., K. Zhang, X. Xu, H. Shen, X. Zhu, Y. Zhang, Y. Hu, and G. Shen. 2020. Substantial changes in nitrogen dioxide and ozone after excluding meteorological impacts during the COVID-19 outbreak in mainland China. *Environmental Science & Technology Letters* 7(6):402-408. https://doi.org/10.1021/acs.estlett.0c00304.

Zheng, B., G. Geng, P. Ciais, S. J. Davis, R. V. Martin, J. Meng, N. Wu, F. Chevallier, G. Broquet, F. Boersma, R. van der A, J. Lin, D. Guan, Y. Lei, K. He, and Q. Zhang. 2020. Satellite-based estimates of decline and rebound in China's CO_2 emissions during COVID-19 pandemic. *Science Advances* 6(49):eabd4998. https://doi.org/10.1126/sciadv.abd4998.

Zheng, B., Q. Zhang, G. Geng, C. Chen, Q. Shi, M. Cui, Y. Lei, and K. He. 2021. Changes in China's anthropogenic emissions and air quality during the COVID-19 pandemic in 2020. *Earth System Science Data* 13(6):2895-2907. https://doi.org/10.5194/essd-13-2895-2021.

Zhou, X., S. Yoon, S. Mara, M. Falk, T. Kuwayama, T. Tran, L. Cheadle, J. Nyarady, B. Croes, E. Scheehle, J. D. Herner, and A. Vijayan. 2021. Mobile sampling of methane emissions from natural gas well pads in California. *Atmospheric Environment* 244:117930. https://doi.org/10.1016/j.atmosenv.2020.117930.

Appendix A

Acronyms, Initialisms, and Glossary

ABLH	atmospheric boundary layer height
AD	activity data
AFOLU	agriculture, forestry, and other land use
API	American Petroleum Institute
AVIRIS-NG	Airborne Visible-Infrared Imaging Spectrometer—Next Generation
BC	black carbon
BUR	biennial update report
CAIT	Climate Analysis Indicators Tool
CARIBIC	Civil Aircraft for the Regular Investigation of the atmosphere Based on an Instrument Container
CBIT	Capacity-building Initiative for Transparency
CDIAC	Carbon Dioxide Information Analysis Center
CEDS	Community Emissions Data System
CEMS	continuous emission monitoring system
CEOS	Committee on Earth Observation Satellites
CFC	chlorofluorocarbon
Climate TRACE	Tracking Real-time Atmospheric Carbon Emissions
CMS	Carbon Monitoring System
CO2M	Copernicus Carbon Dioxide Monitoring mission
COCCON	COllaborative Carbon Column Observing Network
COP	Conference of Parties
DAO	decentralized autonomous organization
EC	eddy covariance
EDF	Environmental Defense Fund

EDGAR	Emissions Database for Global Atmospheric Research
EF	emission factor
EIA	U.S. Energy Information Administration
EO	Earth observation
EPA	U.S. Environmental Protection Agency
ESA	European Space Agency
ESG	environmental, social, and governance
ETF	Enhanced Transparency Framework
EU	European Union
FAIR	findability, accessibility, interoperability, and reusability
FAO	Food and Agriculture Organization of the United Nations
$FFCO_2$	fossil fuel CO_2
FTIR	Fourier transform spectroscopy
GAW	Global Atmospheric Watch
GCoM	Global Covenant of Mayors
GCP	Global Carbon Project
GDP	Gross Domestic Product
GEF	Global Environment Facility
GHG	greenhouse gas
GHGRP	U.S. EPA Greenhouse Gas Reporting Program
GOSAT	Greenhouse Gases Observing Satellite
GRACED	Global Gridded Daily CO_2 Emissions Dataset
GST	Global Stocktake of the Paris Agreement
GWP	global warming potential
HCFC	hydrochlorofluorocarbon
HFC	hydrofluorocarbon
IAGOS	In service Aircraft for a Global Observation System
ICAO	International Civil Aviation Organization
ICLEI	International Council for Local Environmental Initiatives
ICOS	Integrated Carbon Observing System
IEA	International Energy Agency
IG^3IS	WMO Integrated Global Greenhouse Gas Information System
IMEO	International Methane Emissions Observatory
INFLUX	Indianapolis Flux Experiment
InTEM	Inversion Technique for Emission Modelling
IoT	Internet of Things
IPCC	Intergovernmental Panel on Climate Change
ISO	International Organization for Standardization
JAXA	Japan Aerospace Exploration Agency
LCA	life-cycle analysis
LUC	land-use change
LULUC	land use and land-use change
LULUCF	land-use, land-use change, and forestry

APPENDIX A

MAMAP	Methane airborne MAPper
MEIC-HR	Multiresolution Emission Inventory for China—High Resolution
ML	machine learning
NASA	U.S. National Aeronautics and Space Administration
NC	national communication
NDC	nationally determined contribution
NIES	National Institute for Environmental Studies
NIR	national GHG inventory report
NMVOC	non-methane volatile organic compound
NOAA	U.S. National Oceanic and Atmospheric Administration
OC	organic compound
OCO	Orbiting Carbon Observatory
ODIAC	Open-source Data Inventory for Anthropogenic CO_2
OECD	Organisation for Economic Co-operation and Development
OP-FTIR	open path Fourier transform spectroscopy
PermianMAP	Permian Methane Analysis Project
PFC	perfluorocarbon
PM	particulate matter
PRIMAP-hist	Potsdam Real-time Integrated Model for probabilistic Assessment of emissions Paths
PRISMA	PRecursore IperSpettrale della Missione Applicativa
QA/QC	quality assurance/quality control
ROW	rest of world
SCIAMACHY	SCanning Imaging Absorption spectroMeter for Atmospheric CHartographY
SDI	Spatial Data Infrastructure
SLCF	short-lived climate forcer
TCCON	Total Carbon Column Observing Network
TCFD	Task Force for Climate-related Financial Disclosures
TROPOMI	TROPOspheric Monitoring Instrument
U.K.	United Kingdom
UN	United Nations
UNFCCC	United Nations Framework Convention on Climate Change
UNSD	United Nations Statistics Division
U.S.	United States
VOC	volatile organic compound
WDCGG	World Data Centre for Greenhouse Gases
WMO	World Meteorological Organization
WRI	World Resources Institute

Activity-based (or bottom-up) approach: A bottom-up approach begins with details and works up to the highest conceptual level during a given time period, i.e., the individual technologies leading to emissions. The activity-based approach (typically used for emissions inventory development) involves measuring and/or modeling emissions at the scale of individual emitters, such as emitting equipment within factories, power plants, vehicles, landfills, and then extrapolating those results to similar kinds of sources on regional and national scales. This approach involves use of emission factors, a wide spectrum of activity data, and process-based models. In the simplest form, the emission factor is scaled by the corresponding activity data to estimate emissions. However, it can also include direct flux monitoring, ecosystem modeling, and pollution ratio approaches of varying complexity.

Activity data (AD): Data on the magnitude of a human activity resulting in emissions or removals taking place during a given time period. Data on energy use, population, equipment count, metal production, land area, traffic data, lime and fertilizer use, and waste arisings are examples of activity data (IPCC, 2019a).

Aerosol: A suspension of airborne solid or liquid particles, with a typical size between a few nanometers and 10 μm that reside in the atmosphere for at least several hours. The term aerosol, which includes both the particles and the suspending gas, is often used in its plural form to mean aerosol particles. Aerosols may be of either natural or anthropogenic origin. Aerosols may influence climate in several ways: through both interactions that scatter and/or absorb radiation and through interactions with cloud microphysics and other cloud properties, or upon deposition on snow- or ice-covered surfaces thereby altering their albedo and contributing to climate feedback. Atmospheric aerosols, whether natural or anthropogenic, originate from two different pathways: emissions of primary particulate matter (PM), and formation of secondary PM from gaseous precursors. The bulk of aerosols are of natural origin. Scientists often use group labels that refer to the chemical composition, namely: sea salt, organic carbon, black carbon, mineral species (mainly desert dust), sulfate, nitrate, and ammonium. These labels are, however, imperfect as aerosols combine particles to create complex mixtures (IPCC, 2019b).

Agriculture, forestry, and other land use (AFOLU): Agriculture, forestry, and other land use plays a central role for food security and sustainable development. The main mitigation options within AFOLU involve one or more of three strategies: prevention of emissions to the atmosphere by conserving existing carbon pools in soils or vegetation or by reducing emissions of methane and N_2O; sequestration—increasing the size of existing carbon pools, and thereby extracting CO_2 from the atmosphere; and substitution—substituting biological products for fossil fuels or energy-intensive products, thereby reducing CO_2 emissions. Demand-side measures (e.g., by reducing losses and wastes of food, changes in human diet, or changes in wood consumption) may also play a role. FOLU (forestry and other land use)—also referred to as LULUCF (land use, land-use change, and forestry)—is the subset of AFOLU emissions and removals of greenhouse gases resulting from direct human-induced land use, land-use change, and forestry activities excluding agricultural emissions (Allwood et al., 2014).

Air quality: Generally referred to as air pollution that can have negative effects on human health or plant productivity or built environment due to the introduction into the atmosphere of substances (i.e., gases, aerosols) that have a direct or indirect harmful effect (IPCC, 2021).

Annex I Parties/countries: The group of countries listed in Annex I to the United Nations Framework Convention on Climate Change (UNFCCC). Under Articles 4.2 (a) and 4.2 (b) of the UNFCCC, Annex I Parties were committed to adopting national policies and measures with the non-legally binding aim to return their greenhouse gas emissions to 1990 levels by 2000. The group is largely similar to the Annex B Parties to the Kyoto Protocol that also adopted emissions reduction targets for 2008–2012. By default, the other countries are referred to as Non-Annex I Parties (Allwood et al., 2014).

Anthropogenic emissions: Emissions resulting from human activities of greenhouse gases (GHGs), precursors of GHGs, and aerosols. Building on the definition in IPCC AR6 WGI, these activities include emissions associated with the use of fossil fuels (e.g., extraction, processing, transport, and combustion); industrial processes; and agriculture, forestry, and other land use (AFOLU), including deforestation, land-use and land use changes (LULUC), livestock production, fertilization, as well as emissions from agricultural, industrial, and municipal waste management.

Atmospheric-based (or top-down) approach: A top-down approach to a problem is a situation that begins at the highest conceptual level and works down to the details. The atmospheric-based approach estimates emissions using observations of atmospheric concentrations (e.g., ground stations, tall towers, aircraft, and satellites), typically with models that account for atmospheric transport from the emitter to an observation location.

Biennial Update Report (BUR): Reports to be submitted by non-Annex I Parties, containing updates of national greenhouse gas (GHG) inventories, including a national inventory report and information on mitigation actions, needs and support received. Such reports provide updates on actions undertaken by a Party to implement the Convention, including the status of its GHG emissions and removals by sinks, as well as on the actions to reduce emissions or enhance sinks.

Big data: Complex datasets that contain great variety and volume.

Biomass: (1) Includes above- and below-ground living biomass. (2) Organic matter consisting of or recently derived from living organisms (especially regarded as fuel) excluding peat. Includes products, by-products, and waste derived from such material (IPCC, 2019a).

Biomass burning: Biomass burning is the burning of living and dead vegetation (Allwood et al., 2014).

Capacity: Ability of governments to collect and analyze their own emissions data or utilize datasets generated from other sources to inform action.

Capacity-building Initiative for Transparency (CBIT): Initiative that provides support to developing countries in development of its commitments under the ETF, including development of national inventories.

Carbon dioxide (CO_2): The main anthropogenic sources of CO_2 are as a by-product of burning fossil fuels (such as oil, natural gas, and coal), burning biomass, land use changes and industrial processes (e.g., cement production). It is the principal anthropogenic greenhouse gas that affects the Earth's radiative balance (IPCC, 2019a).

Carbon dioxide equivalent (CO_2eq): A metric used to compare the relative warming effect of emissions of various greenhouse gases to CO_2. It is the mass of carbon dioxide that would produce the same estimated radiative forcing as a given mass of another greenhouse gas over a given time period. Carbon dioxide equivalents are computed by multiplying the mass of the gas emitted by its Global warming potential.

Carbon footprint: Total amount of greenhouse gases emitted by an actor or activity during a given period.

Carbon intensity: The amount of emissions of carbon dioxide released per unit of another variable such as gross domestic product, output energy use, or transport (Allwood et al., 2014).

Consumption-based accounting: Consumption-based accounting provides a measure of emissions released to the atmosphere in order to generate the goods and services consumed by a certain entity (e.g., person, firm, country, or region) (Allwood et al., 2014).

Continuous emissions monitoring system (CEMS): The total equipment necessary for the determination of a gas or particulate matter concentration or emission rate using pollutant analyzer measurements and a conversion equation, graph, or computer program to produce results in units of the applicable emission limitation or standard.

Emission factor: A coefficient that quantifies the emissions or removals of a gas per unit activity. Emission factors are often based on a sample of measurement data, averaged to develop a representative rate of emission for a given activity level under a given set of operating conditions.

Emissions: Sources and sinks affecting the various greenhouse gases (i.e., their emissions and removal processes).

Emissions inventory: A set of estimates of the amount of given pollutants or pollutants emitted into the atmosphere from major mobile, stationary, area-wide, and natural source categories over a specific time period such as a day or a year.

Enhanced Transparency Framework (ETF): Addition to the Paris Agreement that guides countries on reporting their greenhouse gas emissions, progress toward their nationally determined contributions, climate change impacts and adaptation, support provided and mobilized, and support needed and received.

Environmental, social, and governance (ESG) metrics: Performance measures or indicators of performance on environmental, social, and governance issues.

Flaring: All burning of natural gas/vapor streams and hydrocarbon liquids by flares as a waste disposal option rather than for the production of useful heat or power (IPCC, 2019a).

Fugitive emissions (oil and natural gas systems): The intentional or unintentional release of greenhouse gases that occurs during the exploration, processing, and delivery of fossil fuels to the point of final use. This excludes greenhouse gas emissions from fuel combustion for the production of useful heat or power. It encompasses venting, flaring, and leaks (IPCC, 2019a).

Global Environment Facility (GEF): Organization that supports the operationalization of the CBIT supporting 72 countries, including least developed and small island developing states.

Global North: An umbrella term used to refer to countries with high relative power and wealth, generally concentrated in the Northern Hemisphere, but also includes Singapore, Japan, South Korea, Australia, and New Zealand.

Global South: Term used to identify regions within Latin America, Asia, Africa, and Oceania that are considered to have relatively low income and often politically or culturally marginalized.

Global Stocktake: Process for taking stock of the implementation of the Paris Agreement with the aim to assess the world's collective progress toward achieving the purpose of the agreement and its long-term goals (UNFCCC, n.d.-a).

Good practice: "Good practice" is a key concept for inventory compilers to follow in preparing national greenhouse gas inventories. The key concept does not change in the IPCC 2019 Refinement. The term "good practice" has been defined since 2000, when this concept was introduced, as "a set of procedures intended to ensure that greenhouse gas inventories are accurate in the sense that they are systematically neither over- nor under-estimates so far as can be judged, and that uncertainties are reduced so far as practicable." This definition has gained general acceptance among countries as the basis for inventory development and its centrality has been retained for the 2019 Refinement. Certain terms in the definition have been updated based on feedback from the statistics community, such that this definition can be also understood as "a set of procedures intended to ensure that greenhouse gas inventories are accurate in the sense that they are systematically neither over- nor underestimates so far as can be judged, and that they are precise so far as practicable" in the context of refinement of Chapter 3 of Volume 13. Good practice covers choice of estimation methods appropriate to national circumstances, quality assurance and quality control at the national level, quantification of uncertainties, and data archiving and reporting to promote transparency (IPCC, 2019a).

Greenhouse gases (GHGs): Gaseous constituents of the atmosphere, both natural and anthropogenic, that absorb and emit radiation at specific wavelengths within the spectrum of radiation emitted by the Earth's surface, by the atmosphere itself, and by clouds. This property causes the greenhouse effect. Water vapor, CO_2, methane, N_2O, and ozone are the primary GHGs in the Earth's atmosphere (IPCC, 2022a). Of these, human emissions directly affect the atmospheric concentrations of CO_2, methane, and N_2O. GHGs occurring in the atmosphere mainly because of human activities include sulfur hexafluoride, hydrofluorocarbons, hydrochlorofluorocarbons, chlorofluorocarbons, halons, and perfluorocarbons; several of these also can cause stratospheric ozone depletion (and their production and consumption are regulated under the Montreal Protocol). Short-lived atmospheric gases with major human emissions like carbon monoxide, nitrogen oxides, and non-methane volatile organic compounds are GHG precursors because they can also increase levels of tropospheric ozone.

Halocarbons: A collective term for the group of partially halogenated species, which includes the chlorofluorocarbons (CFCs), hydrochlorofluorocarbons (HCFCs), hydrofluorocarbons (HFCs), halons, methyl chloride, and methyl bromide. Many of the halocarbons have large global warming potentials (IPCC, 2019a).

Hybrid approach: The hybrid approach derives GHG emissions information through the combination and more complete integration within and between activity-based and atmospheric-based.

In situ measurements: Instrumentation measurements are located directly at the point of interest.

Inverse model: A numerical model for atmospheric processes in which observations are used to infer the values of the parameters characterizing the system under investigation. In top-down analyses, inverse models are used to infer sources and sinks for a greenhouse gas from measurements of the atmospheric or oceanic abundance of that gas.

Isotopic analysis: Identification of abundance of certain stable isotopes of a chemical element.

Land use (change, direct and indirect): Land use refers to the total of arrangements, activities, and inputs undertaken in a certain land cover type (a set of human actions). The term land use is also used in the sense of the social and economic purposes for which land is managed (e.g., grazing, timber extraction, and conservation). In urban settlements it is related to land uses within cities and their hinterlands. Urban land use has implications on city management, structure, and form and thus on energy demand, greenhouse gas emissions, and mobility, among other aspects (Allwood et al., 2014).

Land-use change (LUC): Land-use change refers to a change in the use or management of land by humans, which may lead to a change in land cover. Land cover and LUC may have an impact on the surface albedo, evapotranspiration, sources and sinks of GHGs, or other properties of the climate system and may thus give rise to radiative forcing and/or other impacts on climate, locally or globally (IPCC, 2000).

Indirect land-use change (iLUC): Indirect land-use change refers to shifts in land use induced by a change in the production level of an agricultural product elsewhere, often mediated by markets or driven by policies. For example, if agricultural land is diverted to fuel production, forest clearance may occur elsewhere to replace the former agricultural production (Allwood et al., 2014).

Land use, land-use change, and forestry (LULUCF): A greenhouse gas inventory sector that covers emissions and removals of GHGs resulting from direct human-induced land use, land-use change, and forestry activities excluding agricultural emissions. See also Agriculture, Forestry and Other Land Use (AFOLU) (Allwood et al., 2014).

Least developed countries (LDCs): According to the UN, low-income countries confronting severe structural impediments to sustainable development. They are highly vulnerable to economic and environmental shocks and have low levels of human assets.

Life-cycle analysis (LCA): A cradle-to-grave or cradle-to-cradle analysis technique to assess environmental impacts associated with all the stages of a product's life, which is from raw material extraction through materials processing, manufacture, distribution, and use (Muralikrishna and Manickam, 2017).

Managed land (vegetation): Land area where human interventions and practices have been applied to perform production, ecological or social functions (IPCC, 2019a).

Methane (CH_4): The main anthropogenic sources of methane, a greenhouse gas, are from three sectors: energy production from fossil fuels, agriculture, and waste (Saunois et al., 2020). Energy production accounts for about 35 percent of anthropogenic methane emissions, agriculture accounts for about 40 percent, and waste accounts for about 20 percent (Saunois et al., 2020; UNEP and CCAC, 2021). Methane is the major component of natural gas and associated with all fossil-based hydrocarbon fuels. Significant anthropogenic emissions also occur as a result of animal husbandry and paddy rice production. Methane is also produced where organic matter decays under anaerobic conditions (IPCC, 2019a), with anthropogenic emissions from landfill waste and wastewater systems. Approximately 35–50 percent of annual methane emissions are from natural sources such as wetlands (Saunois et al., 2020). Under future global warming, there is potential for increased methane emissions from thawing permafrost, wetlands, and subsea gas hydrates.

Mitigation: In the context of climate change, mitigation relates to a human intervention to reduce the sources or enhance the sinks of greenhouse gases. Examples include using fossil fuels more efficiently for industrial processes or electricity generation, switching to solar energy or wind power, improving the insulation of buildings and expanding forests and other "sinks" to remove greater amounts of CO_2 from the atmosphere.

National GHG Inventory Report (NIR): A report containing transparent and detailed information on the inventory. It should include descriptions of the methodologies used in the estimations (including references and sources of information), the data sources, the institutional arrangements for the preparation of the inventory (including quality assurance and control procedures), and recalculations and changes compared with the previous inventory.

Nationally determined contribution (NDC): Submissions by countries that have ratified the Paris Agreement which presents their national efforts to reach the Paris Agreement's long-term temperature goal of limiting warming to well below 2°C. New or updated NDCs were expected to be submitted in 2020 and should be submitted every 5 years thereafter. NDCs thus represent a country's current ambition/target for reducing emissions nationally.

Nitrous oxide (N_2O): The main anthropogenic source of N_2O, a greenhouse gas, is agriculture (fertilizer, soil and animal manure management), but important contributions also come from wastewater treatment, fossil fuel combustion, and chemical industrial processes (Allwood et al., 2014). N_2O is also produced naturally from a wide variety of biological sources in soil and water, particularly microbial action in wet tropical forests (Allwood et al., 2014).

Non-Annex I Parties/countries: Non-Annex I Parties are mostly developing countries. Certain groups of developing countries are recognized by the United Nations Framework Convention on Climate Change (Convention) as being especially vulnerable to the adverse impacts of climate change, including countries with low-lying coastal areas and those prone to desertification and drought. Others, such as countries that rely heavily on income from fossil fuel production and commerce, feel more vulnerable to the potential economic impacts of climate change response measures. The Convention emphasizes activities that promise to answer the special needs and concerns of these vulnerable countries, such as investment, insurance, and technology transfer. See also Annex I Parties/countries (Allwood et al., 2014).

Ozone (O_3): The triatomic form of oxygen, and a gaseous atmospheric constituent. In the troposphere, ozone is created both naturally and by photochemical reactions involving gases resulting from human activities (e.g., smog). Tropospheric ozone acts as a greenhouse gas. In the stratosphere, ozone is created by the interaction between solar ultraviolet radiation and molecular oxygen (O_2). Stratospheric ozone plays a dominant role in the stratospheric radiative balance. Its concentration is highest in the ozone layer.

Particulate matter (PM): Very small particles emitted during the combustion of biomass and fossil fuels or produced through atmospheric chemical interactions. PM may consist of a wide variety of substances. Of greatest concern for health are particulates of diameter less than or equal to 10 μm, usually designated as PM_{10} (Allwood et al., 2014), especially those less than 2.5 μm ($PM_{2.5}$). Also see Aerosols.

Point source: Any single identifiable source of pollution from which pollutants are discharged, such as a pipe, ditch, ship, or factory smokestack.

Precursors: Atmospheric compounds that are not greenhouse gases (GHGs) or aerosols, but that influence GHG or aerosol concentrations by taking part in physical or chemical processes regulating their production or destruction rates (IPCC, 2019a).

Production inventories: Inventories that tie emissions to the geographic location where they enter the atmosphere. This is often also referred to as "Scope 1" emissions, a term that emanates from the scope language routinely used at subnational spatial scales.

Radiative forcing: The change in the net, downward minus upward, radiative flux (expressed in $W\ m^{-2}$) due to a change in an external driver of climate change, such as a change in the concentration of CO_2, the concentration of volcanic aerosols, or in the output of the Sun. The stratospherically adjusted radiative forcing is computed with all tropospheric properties held fixed at their unperturbed values, and after allowing for stratospheric temperatures, if perturbed, to readjust to radiative-dynamical equilibrium. Radiative forcing is called instantaneous if no change in stratospheric temperature is accounted for. The radiative forcing once both stratospheric and tropospheric adjustments are accounted for is termed the effective radiative forcing.

Removals: Removal of greenhouse gases and/or their precursors from the atmosphere by a sink (IPCC, 2019a).

Scope 1, Scope 2, and Scope 3 emissions: Emissions responsibility as defined by the GHG Protocol, a private sector initiative. "Scope 1" indicates direct greenhouse gas (GHG) emissions that are from sources owned or controlled by the reporting entity. "Scope 2" indicates indirect GHG emissions associated with the production of electricity, heat, or steam purchased by the reporting entity. "Scope 3" indicates all other indirect emissions, i.e., emissions associated with the extraction and production of purchased materials, fuels, and services, including transport in vehicles not owned or controlled by the reporting entity, outsourced activities, waste disposal, etc. (WRI/WBCSD, 2004).

Sector: An emission-producing segment of the economy such as energy; industrial processes and product use; agriculture, forestry, and other land use; and waste (IPCC, 2019a).

Short-lived climate forcers (SLCFs): Following the IPCC definition, chemically reactive compounds with short (relative to CO_2) atmospheric lifetimes (from hours to about two decades) but characterized by different physiochemical properties and environmental effects. Their emission or formation has a significant effect on radiative forcing over a period determined by their respective atmospheric lifetimes. Changes in their emissions can also induce long-term climate effects via, in particular, their interactions with some biogeochemical cycles. SLCFs are classified as direct or indirect, with direct SLCFs exerting climate effects through their radiative forcing and indirect SLCFs being the precursors of other direct climate forcers. Direct SLCFs include methane, ozone, primary aerosols, and some halogenated species. Indirect SLCFs are precursors of ozone or secondary aerosols. SLCFs can be cooling or warming through interactions with radiation and clouds. They are also referred to as near-term climate forcers. Many SLCFs are also air pollutants. A related term "short-lived climate pollutant" is used to denote the set of climate forcers that are warming (methane, black carbon, hydrofluorocarbons, and ozone) and avoid bundling with cooling aerosols (primarily sulfates and nitrates) associated with fossil fuel combustion. Such bundling can lead to coincidental canceling of forcing that leads to the inaccurate perception that emissions of these short-lived species are independent from policies affecting CO_2 emissions from fossil fuel usage.

Sink: Any process, activity or mechanism that removes a greenhouse gas, an aerosol, or a precursor of a greenhouse gas from the atmosphere (IPCC, 2019a).

Small Island Developing States (SIDS): A distinct group of 38 Member States and 20 Non-UN Members/Associate Members of United Nations regional commissions that face unique social, economic, and environmental vulnerabilities.

Source: Any process, activity, or mechanism that releases a greenhouse gas, an aerosol, or a precursor of a greenhouse gas or aerosol into the atmosphere. Certain activities, such as forestry, can be both a source and a sink of greenhouse gas emissions.

Surrogate (proxy) data: Surrogate data, or so-called proxy data, are data that are used in place of the actual data, where the specific data needed are unobtainable. Often surrogate data are needed to describe changes in an emission source over time, for example population change may be used to approximate change in waste arisings (IPCC, 2019a).

Tier: A tier represents a level of methodological complexity. Usually three tiers are provided. Tier 1 is the basic method, Tier 2 intermediate, and Tier 3 most demanding in terms of complexity and data requirements. Tiers 2 and 3 are sometimes referred to as higher tier methods and are generally considered to be more accurate (IPCC, 2019a).

United Nations Framework Convention on Climate Change (UNFCCC): The Convention was adopted on 9 May 1992 in New York and signed at the 1992 Earth Summit in Rio de Janeiro by more than 150 countries and the European Community. Its ultimate objective is the "stabilisation of greenhouse gas concentrations in the atmosphere at a level that would prevent dangerous anthropogenic interference with the climate system." It contains commitments for all Parties under the principle of "common but differentiated responsibilities." Under the Convention, Parties included in Annex I aimed to return greenhouse gas emissions not controlled by the Montreal Protocol to 1990 levels by the year 2000. The convention entered in force in March 1994. In 1997, the UNFCCC adopted the Kyoto Protocol (Allwood et al., 2014).

Validation: Validation is the establishment of sound approach and foundation. In the context of emissions inventories, validation involves checking to ensure that the inventory has been compiled correctly in line with reporting instructions and guidelines. It checks the internal consistency of the inventory. The legal use of validation is to give an official confirmation or approval of an act or product (IPCC, 2019a).

Verification: An independent examination of a greenhouse has inventory to help establish whether the actual emissions are consistent with available observations and other analyses.

Appendix B

Atmospheric Observations: Methods and Examples

Atmospheric greenhouse gas (GHG) measurements are collected with a variety of techniques, including ground, aircraft, and satellite based, in order to capture the range of characteristics of GHGs. Some gases have very small horizontal and vertical gradients (i.e., CO_2), while others, which react more quickly in the atmosphere (e.g., methane), have substantial atmospheric gradients, both horizontally and vertically. For a gas like CO_2, on the one hand, small spatial gradients make it possible to generally characterize the global distribution with a small number of measurements across the globe. On the other hand, the fluxes of interest (e.g., changes due to a specific emission sector or source) create only small gradients, so it is critical that measurement techniques can sense small concentration gradients, especially in areas where emission and removal processes of interest are occurring. In order to observe GHGs across a range of spatial and temporal scales, a range of measurement techniques are used (Figure B-1) and described in detail in the following section.

Surface-based observations

In situ (i.e., continuous) ground measurements of atmospheric GHGs started with CO_2 in 1958 at the Mauna Loa observatory, Hawaii. This dataset, known as the reference CO_2 time series or "Keeling Curve" (Keeling et al., 1976), helped to determine CO_2 growth rate and trends and the rapid increase of global atmospheric CO_2. This increase is attributed to the accumulation of about half of anthropogenic CO_2 emissions in the atmosphere (Friedlingstein et al., 2019). Since then, many in situ measurement stations have been developed to improve understanding of the variability and trends of GHG emissions and carbon sinks at different scales. Information about global and regional measurement networks is managed by the World Data Centre for Greenhouse Gases (WDCGG) program.[1] The World Meteorological Organization (WMO) is responsible for international GHG calibration scales, which are deployed on all stations reporting to the WDCGG

---

[1] https://gaw.kishou.go.jp/

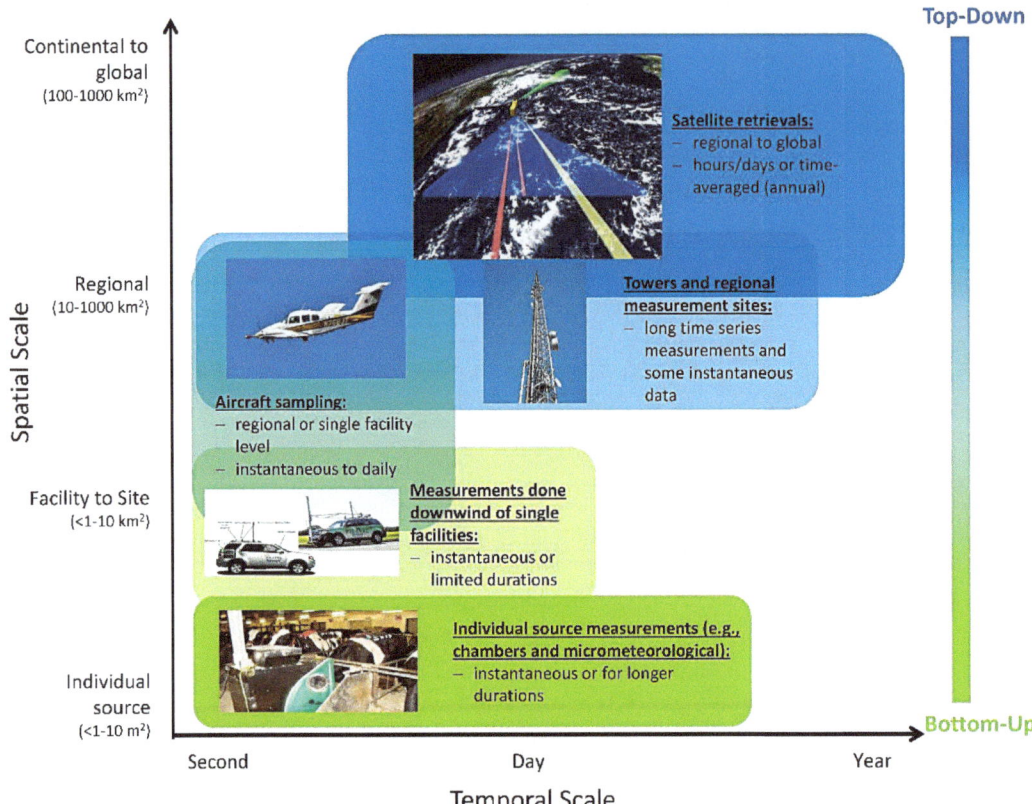

FIGURE B-1 Greenhouse gas measurement platforms across a range of spatial (vertical axis) and temporal scales (horizontal axis). The colored bar on the right represents the spectrum of analytical methods to analyze these measurements from bottom-up or activity-based (green) to top-down or atmospheric-based (blue).
SOURCES: NASEM, 2018; NASA; NOAA; Paul Shepson, Stony Brook University; Tsai et al., 2017; and Alexander Hristov, Pennsylvania State University.

program. Most stations measure in situ CO_2, methane (CH_4), and N_2O, and, in some cases, monitor SF_6 and F gases.

The global in situ GHG network is currently composed of 32 stations located in remote locations to monitor the representative large-scale trends not influenced by major local sources of atmospheric CO_2, CH_4, N_2O, SF_6, and fluorinated gases in both hemispheres, and to infer information on the latitudinal distribution of sources and sinks.[2] Most stations are equipped with meteorological sensors and some are also instrumented to monitor emission tracers (e.g., Radon-222, carbon isotopes, carbon monoxide [CO], oxygen) used as tools for partitioning anthropogenic versus natural fluxes (Ciais et al., 1995; Tans et al., 1990) and separating oceanic versus continental fluxes (Battle et al., 2000; Keeling et al., 1996). The U.S. National Oceanic and Atmospheric Administration (NOAA) deployed stations on a north-to-south gradient at four baseline observatories: Utqiaġvik (formerly Barrow), Alaska; Mauna Loa, Hawaii; American Samoa; and South Pole, Antarctica.[3]

---

[2] The initial sites were placed in remote locations far from anthropogenic emissions sources to provide a more accurate measure of the trends in well-mixed, long-lived GHGs. This siting, however, limits the ability of these stations to precisely identify specific sources of emissions, as was found in the case of "unexpected" CFC-11 emissions (Montzka et al., 2018).

[3] https://gml.noaa.gov/ccgg/insitu/#:~:text=NOAA%20Baseline%20Observatories&text=GML%20measures%20 greenhouse %20gases%20at,CO%20measurements%20in%20the%201980's

Other countries then joined the global in situ GHG network including Canada, China, Japan, Australia, and various European countries. The global CO_2 datasets are used to constrain global inversion models (e.g., Bousquet et al., 2000; Denning et al., 1995; Fernández-Martínez et al., 2019; Rayner et al., 2008) and assimilation systems such as CarbonTracker,[4] Copernicus Atmosphere Monitoring Service (e.g., Pinty et al., 2019), and the National Aeronautics and Space Administration's (NASA's) Goddard Earth Observing System (GEOS) (Weir et al., 2021).

Continental GHG networks. Continental networks have been developed over the last two decades to better assess trends and variability of atmospheric GHG concentrations, sources, and sinks at the regional to continental scales. Continental networks are also used to detect trends in atmospheric GHGs (e.g., Ramonet et al., 2010) to constrain inverse modeling frameworks (e.g., Monteil et al., 2020), and to study the impact of climate events on natural ecosystems, such as drought events (e.g., Smith et al., 2020). These networks monitor atmospheric GHGs in the lower atmosphere (i.e., continental boundary layer and the free troposphere) with stations deployed on tall towers or on top of mountains and far from local sources to ensure a regional representativeness of each site.[5] The sites are equipped with meteorological sensors, which provide key ancillary datasets such as wind speed and direction to analyze atmospheric GHG variability (Yver-Kwok et al., 2021). Some sites are also equipped with remote sensing instruments (e.g., lidar, ceilometers) and make measurements of other gases (e.g., CO) that can be used to identify anthropogenic versus natural fluxes (e.g., Palmer et al., 2006).

In Europe, the GHG continental in situ network, ICOS-Atmosphere (Integrated Carbon Observing System[6]), has been operational since 2012 and includes about 50 in situ stations. Each country operates its own network and the data are sent every day and integrated into one central database with harmonized protocols (Hazan et al., 2016). The datasets feed the European Copernicus GHG assimilation system.[7] In the United States, the continental network is managed by NOAA and started in the 1990s.[8] All sites are equipped with tall towers with standardized instrumentation, and a central datacenter, and are linked to the international WMO GHG calibration scales. The NOAA tall tower network is part of the North American Carbon Program and its datasets used as a primary input for NOAA's Carbon Tracker CO_2 and methane data assimilation systems.

National GHG networks. National GHG networks are partly embedded into continental networks, for example in Europe and in the United States, and used to assess GHG emissions at the national scale. Several studies assessed emissions inventories independently through atmospheric observation-based methods, as detailed in Chapter 2.

Urban GHG networks. Since the 2010s, urban GHG networks have been developed to better quantify emissions and understand uncertainty at these scales. Such measurements address mostly CO_2 across all sectors; although urban emissions are often dominated by energy consumption in the building and transportation sectors, some "producer" cities have considerable emissions in the industrial sector. Efficient urban measurements are designed with sites deployed in both dominant upwind and in-path or downwind locations with background regional sites upwind of the city and sites downwind of the city. Urban atmospheric GHG enhancements are then determined by subtracting the data collected at the background sites from the data collected at the urban sites. Additional measurements of emission tracers such as CO, nitrogen oxides (NO_x), volatile organic compounds (VOCs), and carbon isotopes can be key tools to partition fossil vs modern fluxes and to discrimi-

---

[4] https://gml.noaa.gov/ccgg/carbontracker

[5] https://www.icos-cp.eu/observations/atmosphere/stations

[6] https://www.icos-cp.eu

[7] https://atmosphere.copernicus.eu/ghg-services

[8] https://gml.noaa.gov/ccgg/insitu/#:~:text=NOAA%20Baseline%20Observatories&text=GML%20measures%20 greenhouse%20gases%20at,CO%20measurements%20in%20the%201980's

nate source categories (e.g., Ammoura et al., 2016; Graven et al., 2018). Ancillary meteorological parameters are necessary to interpret atmospheric GHG variability (e.g., Xueref-Remy et al., 2018).

Several intensive urban measurement efforts have been deployed in recent years, starting in 2010 with the INFLUX project in Indianapolis (Turnbull et al., 2015), the MEGACITIES project in the Los Angeles megacity (Verhulst et al., 2017), and the CO_2-MEGAPARIS project in the Paris megacity (Xueref-Remy et al., 2018). Other urban intensives that have been developed since then include the Northeast United States corridor from Boston to Washington, DC (Karion et al., 2020), Salt Lake City (Lin et al., 2018; Mitchell et al., 2018), São Paulo, Brazil (Pérez-Martínez et al., 2020), Tokyo, Japan (Imasu and Tanabe, 2018), Aix-Marseille, France (Xueref-Remy et al., 2020b), and other European cities as a test bed of the ICOS cities project.[9] The WMO Integrated Global Greenhouse Gas Information System (IG^3IS) scientific community recently published a first report on urban network design recommendations.[10] Systemic urban GHG measurements are still rare in the Global South.

Mobile in situ measurements. Mobile in situ measurements (e.g., using a car or train platform) bring additional information to in situ urban measurements, particularly on characterizing GHG spatial variability (Idso et al., 2001; Mallia et al., 2020). Mobile campaigns in urban areas are also used to infer methane sources in cities, mostly from leaking gas lines (Carranza et al., 2022; Phillips et al., 2013; Xueref-Remy et al., 2020a).

Eddy covariance (EC) flux measurements. Eddy covariance is a technique based on the principle of atmospheric turbulence, which measures GHG fluxes (mostly CO_2 and CH_4) from high-frequency GHG concentrations and 3D wind measurements. Measurements are usually done on a mast several meters above the canopy of buildings/vegetation or in an open-view area to avoid perturbations from surrounding structures/trees. The footprint of the EC measurements is usually determined from atmospheric dispersion models or analytical solutions of the diffusion equation (e.g., Kljun et al., 2015). Typically, the footprint of an EC station is a hundred square meters to some square kilometers (Rebmann et al., 2018). While most often deployed on natural areas, the EC approach has recently been used in some cities to assess urban CO_2 and methane emissions, for example in the center of London (Helfter et al., 2016), Helsinki (Järvi et al., 2012), and Florence (Gioli et al., 2015).

EC flux measurement sites have been mostly developed starting in the 1990s to assess local, regional, national, and continental fluxes and were further organized into networks within large and long-term observation infrastructures, for example in Europe (ICOS-Ecosystems[11]) and in the United States (FLUXNET;[12] NEON;[13] AMERIFLUX[14]). Over the last decade, some EC stations were deployed to assess urban CO_2 and CH_4 emissions and to independently verify local emissions inventories (e.g., Helfter et al., 2016). Such measurements were also used to assess the impact of the COVID-19 lockdown on CO_2 emissions. For example, an EC study conducted in 11 European cities estimated a reduction of 5 to 87 percent of urban CO_2 emissions during that period, mostly from a decrease in traffic (Nicolini et al., 2022).

Point-source in situ measurements. Point-source in situ measurements have been performed to improve estimates of anthropogenic GHG sources and emissions. Point-source measurements of methane from mobile campaigns have been used mainly for (1) detecting methane gas leaks (e.g., Keyes et al., 2020); (2) identifying methane point sources (e.g., Zazzeri et al., 2015); and (3)

---

[9] https://www.icos-cp.eu/projects/icos-cities-project#:~:text=ICOS%20Cities%20(PAUL%20%E2%80%93Pilot%20Applications,measurement%20system%20for%20urban%20areas
[10] https://ig3is.wmo.int/en/news/towards-international-standard-urban-ghg-monitoring-and-assessment
[11] https://www.icos-cp.eu/observations/ecosystem/etc
[12] https://fluxnet.org
[13] https://www.neonscience.org/about
[14] https://ameriflux.lbl.gov

quantifying methane source emissions (e.g., Yacovitch et al., 2015). Such campaigns have been widely developed in several countries in the last decade due to the large underestimation of methane emissions. Point-source measurements allow an independent assessment of methane sources given by regional and national inventories and reveal missing or underestimated sources (e.g., Al-Shalan et al., 2022; Marklein et al., 2021; Varon et al., 2019; Xueref-Remy et al., 2020a). Point-source measurements have also been used to assess fossil fuel CO_2 emissions, for example, by sampling transects across an industrial CO_2 source plume stripped from natural gas and vented to the atmosphere in New Zealand (Turnbull et al., 2014).

Ground-based remote sensing. There is a long history of using open path Fourier transform spectroscopy (OP-FTIR) measurements to characterize atmospheric gases, generally at the facility scale with commercially available measurement systems. More recent research has expanded this to use ultraviolet and visible light, and, more recently, near-infrared light (Griffith et al., 2018). Recent innovations led to CO_2 and methane measurements over distances of 1.5 km with improvements in measurement repeatability and reduced bias, extending the application of this technique from the industrial scale to city scale. Specific plumes can be detected, and in some cases quantified, with commercial and thermal infrared (Ravikumar et al., 2017) and in research work with stationary hyperspectral cameras using shortwave infrared (Knapp et al., 2022).

Uplooking FTIR measurements have been deployed in a global validation network utilized by the satellite community. This technique relies on direct sun measurements at (near infrared) wavelengths to characterize to total atmospheric column a wide range of trace gases. The data are used for scientific investigations as well as satellite validation. The Total Carbon Column Observing Network (TCCON) uses a Bruker FTIR, Suntracker, and is typically housed in a shipping container. The TCCON sites are periodically overflown by aircraft or Aircore to establish a connection between the uplooking data and the standards used with in situ measurements. A more portable, lower cost instrumentation approach has been developed using the EM-27 spectrometer (Gisi et al., 2012). These are being deployed in research networks in Berlin (Rißmann et al., 2022), Toronto (Mostafavi Pak et al., 2019), Namibia (Frey et al., 2021), and Uganda (Humpage et al., 2020) and in a new validation network, COCCON (COllaborative Carbon Column Observing Network) (Frey et al., 2019).

Aircraft-based observations

Aircraft in situ measurements. Regular commercial aircrafts have been used as scientific platforms for monitoring atmospheric GHGs in the troposphere and at the upper troposphere/lower stratosphere levels; such programs have been developed in the 1980s in Japan (CONTRAIL; Matsueda et al., 2002, 2008) and then by Germany as the CARIBIC (Civil Aircraft for the Regular Investigation of the atmosphere Based on an Instrument Container) flying observatory,[15] further completed in the 2010s by several European countries within the IAGOS Research Infrastructure.[16] Urban networks do not exist in all cities, and airborne campaigns could be performed more regularly to provide emissions estimates and a regular independent assessment of emission inventories.

Aircraft remote sensing. Aircraft remote sensing has been used widely for CO_2 and methane measurements, while other GHGs (e.g., N_2O, F gases) have not been sampled with aircraft remote sensing techniques. Some recent reports of measurements include the AVIRIS-NG (Airborne Visible-Infrared Imaging Spectrometer—Next Generation), PRISMA (PRecursore IperSpettrale della Missione Applicativa), and GAO (Global Airborne Observatory) sensors for CO_2 and methane measurements (Cusworth et al., 2021a, 2021b). The aircraft instrument MAMAP (Methane

---

[15] https://www.caribic-atmospheric.com/
[16] https://www.iagos.org/

airborne MAPper) has also been developed in support of the CO2M mission development and has been used to measure CO_2 (Krings et al., 2018) and methane (Krautwurst et al., 2021). GOBLEU[17] (Greenhouse gas Observations of Biospheric and Local Emissions from the Upper sky) is an aircraft program jointly launched by Japan Aerospace Exploration Agency (JAXA) and All Nippon Airways (ANA). GOBLUE will collect CO_2 and NO_2 measurements from ANA commercial aircrafts and prototypes the combined use of CO_2 and NO_2 for better emissions monitoring.

Space-based observations

Major advances in satellite remote sensing of GHGs were made in the 2010s and 2020s. The evolution of measurements and roadmap for future missions are well described in reports such as CEOS (2018) and the European Commission's CO_2 Reports (Ciais, 2015; Pinty et al., 2017, 2019) Satellite measurements (Figure B-2) have been used to quantify GHG emissions from power plants, observe enhancements from cities, and estimate fluxes from a wide set of regions of the world, as well as provide global overviews. As detailed in NASEM (2018), satellite measurements have evolved from large (30 km × 60 km for the SCIAMACHY experiment) footprint measurements. Footprint sizes were reduced in satellites that followed, including GOSAT (80 km^2) and then TROPOMI (7 km × 7 km). Satellite measurements continue to improve in spatial and temporal resolution with the next generation of satellites (Figure B-2).

Multigas approaches and isotopic analysis

Isotopic analysis. As radiocarbon (^{14}C) is radioactive with a lifetime of ~5,730 years, ^{14}C measurements in atmospheric CO_2 are used at the urban to the national scales to discriminate CO_2 emissions from the burning of fossil fuels (which do not contain any radiocarbon atoms) from modern CO_2 fluxes (i.e., from vegetation fluxes or wood burning) (Miller et al., 2012). ^{14}C measurements of methane are rare because of the low signal to be detected, but emerging techniques are promising (Zazzeri et al., 2021). The ^{13}C to ^{12}C ratio depends on CO_2 sources and can be used to discriminate different sources of CO_2 and methane through Keeling plot or Tans-Miller plot methods, especially for urban CO_2 (e.g., Lopez et al., 2013; Zazzeri et al., 2021) and point sources—mostly through mobile campaigns—for methane (e.g., Al-Shalan et al., 2022; Xueref-Remy et al., 2020a). Isotopic information can be compared to emission inventories and used to improve source characterization (e.g., completeness, location, intensity). Both ^{14}C and ^{13}C methods have been used at the global scale to quantify the impact of the increasing burden of anthropogenic emissions on atmospheric CO_2 since the pre-industrial era.[18]

Tracer to tracer approach. Most GHG observation programs combine in situ CO_2 measurements with emission tracers (e.g., non-methane volatile organic compounds [NMVOCs], CO, NO_x, black carbon [BC], sulfur oxides [SO_x]) collected by research institutes or air quality agencies. These emission tracers can be used to discriminate between emission sources by analyzing correlations and ratios between CO_2 and the emission tracers. Most combustion processes are incomplete, producing not only CO_2 but other species such as CO, NO_x, VOCs, BC, and SO_x. Some species are typical of some sources—for example, NO_x and isopentane are often attributed to traffic, ethane and butane are more associated with natural gas heating, and SO_x is more associated with combustion of sulfur-containing fossil fuels such as coal in electricity generation and diesel or heavy fuel oils used in transport and shipping. BC mixing ratios peak depending on the absorption wavelength of BC particles and can be attributed either to wood burning or to fossil fuel burning sources.

---

[17] https://www.eorc.jaxa.jp/GOSAT/ANAexp/index_e.html
[18] https://gml.noaa.gov/ccgg/isotopes/c13tellsus.html

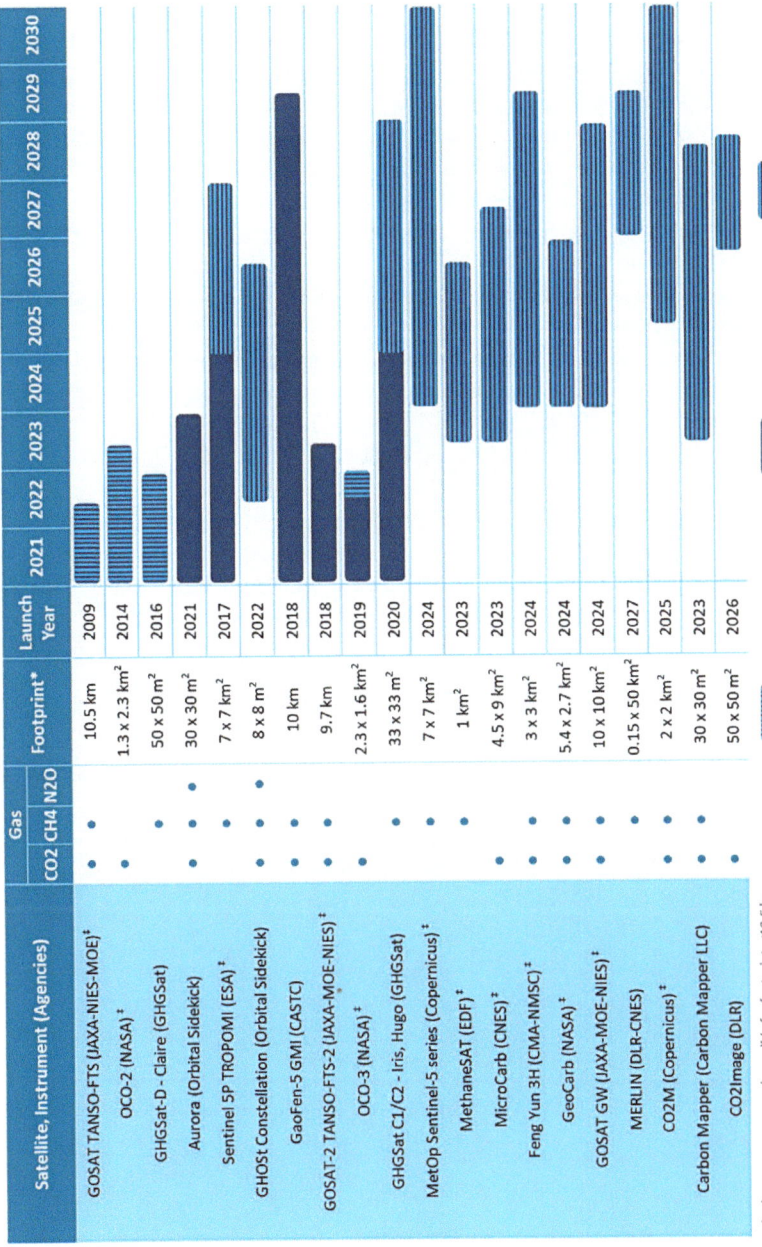

FIGURE B-2 Current and planned satellite missions measuring CO_2, methane (CH_4), and/or N_2O. SOURCE: Adapted from CEOS, 2018.

Analyzing the correlations and ratios between CO_2 and these tracers, especially in concomitant mixing ratio CO_2-tracer peaks, especially at key time periods such as rush hour and domestic heating, is a method often used to identify the different sources of CO_2 and their temporal variability (e.g., Lopez et al., 2013). This method has highlighted inaccurate emission ratios between emission tracers and CO_2 in inventories, improving emission ratios and source partitioning in urban to national inventories (e.g., Ammoura et al., 2016). Methane is also co-emitted with other species, such as ethane, which is one component of natural gas. Ethane-to-methane correlations have been used to estimate fugitive methane emissions from urban centers (Plant et al., 2019) and revealed an underestimation of methane emissions in the inventories of these cities. It is worth noting that using a single value for ethane-to-methane ratios—for example, for the entire oil and gas supply chain—can introduce significant errors into source attribution (Allen, 2016).

Dual tracer approaches can be used to calculate methane emissions at point sources, such as landfills, natural gas storage sites, wastewater treatment facilities, and cattle farms, by releasing a tracer such as acetylene (due to its negligible background level) at the methane source location at a known emission rate. The measured ratio of methane to a tracer downwind of the source plume, combined with the known tracer emission rate, provides a quantification of the methane emission rate and delivers an independent estimate of methane emissions of that source from the inventory. This approach can be used to improve emissions inventories (e.g., Roscioli et al., 2015; Vechi et al., 2022; Zavala-Araiza et al., 2018).

Radon-222 is a natural tracer emitted by the continental crust with a lifetime of 3.8 days, typical of the synoptic scale. Its emission rate varies with the nature of soils and is available in the literature for different continental surfaces. Concomitant atmospheric measurements of radon-222 and GHGs are used with a known Rn-222 flux and the air mass footprint (i.e., the surface seen by the air mass before reaching the measurement site) from wind measurements to calculate GHG emissions at the regional scale. Such results help to assess independent regional and national inventories for species such as CO_2 and N_2O (e.g., Schmidt et al., 2001, 2003; Xueref-Remy et al., 2020b).

Appendix C

Contributors of Input to the Study

Caroline Alden, LongPath Technologies Inc.
David Allen, University of Texas at Austin
Arlyn Andrews, National Oceanic and Atmospheric Administration (NOAA) Global Monitoring Laboratory
Fouzi Benkhelifa, NEXQT
Rostyslav Bun, Lviv Polytechnic National University; WSB University at Dąbrowa Górnicza
Asa Carré-Burritt, Bridger Photonics, Inc.
Abhishek Chatterjee, National Aeronautics and Space Administration (NASA) Jet Propulsion Laboratory
Ronald Cohen, University of California, Berkeley
David Crisp, NASA Jet Propulsion Laboratory; California Institute of Technology (Retired)
Phil DeCola, University of Maryland, College Park
Mausami Desai, U.S. Environmental Protection Agency (EPA)
Riley Duren, Carbon Mapper; University of Arizona; NASA Jet Propulsion Laboratory
Louisa Durkin, Open Earth Foundation
Richard Engelen, European Centre for Medium-Range Weather Forecasts
Michael Gillenwater, Greenhouse Gas Management Institute
Patrick Gonzalez, University of California, Berkley
Deborah Gordon, Rocky Mountain Institute; Brown University
Aarti Gupta, Wageningen University
Dorit Hammerling, Colorado School of Mines
Lisa Hanle, Independent Consultant
Jorn Herner, California Air Resources Board
Jason Hickel, Autonomous University of Barcelona; London School of Economics
Lena Höglund-Isaksson, International Institute for Applied Systems Analysis
Francesca Hopkins, University of California, Riverside

Lei Hu, NOAA Global Monitoring Laboratory
Jisoon Ihm, Pohang University of Science and Technology
Daniel Jacob, Harvard University
Atul Jain, University of Illinois Urbana-Champaign
Greet Janssens-Maenhout, European Commission – Joint Research Centre
Rodrigo Jimenez, Universidad Nacional de Colombia
Elizabeth Joyner, NASA Goddard Space Flight Center
Lynn Kaack, Hertie School
Sekou Keita, University Peleforo Gon Coulibaly
Dong-Gill Kim, Hawassa University
Takeshi Kuramochi, NewClimate Institute; Utrecht University
David Lary, University of Texas at Dallas
Pieternel Levelt, National Center for Atmospheric Research
Wenjie Liao, Sichuan University
John C. Lin, University of Utah
Anders Lindroth, Lund University
Junjie Liu, NASA Jet Propulsion Lab
Zhu Liu, Tsinghua University
Shuaib Lwasa, Global Center on Adaption
Lauren Maffeo, Steampunk
Shamil Maksyutov, National Institute for Environmental Studies (NIES) Japan
Gregg Marland, Appalachian State University
Tsuneo Matsunaga, NIES Japan
Gavin McCormick, WattTime; Climate TRACE
Jason McKeever, GHGSat Inc.
Yasjka Meijer, European Space Agency
Anna Michalak, Stanford University; Carnegie Mellon University
John Miller, NOAA Global Monitoring Laboratory
Jan Minx, Mercator Research Institute on Global Commons and Climate Change
Stephen Montzka, NOAA Global Monitoring Laboratory
Daniel Moran, Norwegian University of Science and Technology
M. Granger Morgan, Carnegie Mellon University
Kim Mueller, National Institute of Standards and Technology
Filomena Nelson, Secretariat of the Pacific Regional Environment Programme
Lesley Ott, NASA Goddard Space Flight Center
Philippe Peylin, Laboratoire des Sciences du Climat et de l'Environnement
Ben Poulter, NASA Goddard Space Flight Center
Jeff Privette, NOAA
Abhilasha Purwar, Blue Sky Analytics
Yann Robiou du Point, Open Earth Foundation
Selina Roman-White, Cheniere Energy Inc.
Steven Smith, Pacific Northwest National Laboratory
Jonathan Stern, Oxford Institute for Energy Studies
Gyami Shrestha, U.S. Global Change Research Program
Ian Tout, United Nations Framework Convention on Climate Change
Jocelyn Turnbull, CarbonWatch New Zealand
Melissa Weitz, U.S. EPA

Molly White, Greenhouse Gas Management Institute
Christine Wiedinmyer, Cooperative Institute for Research in Environmental Sciences; University of Colorado Boulder
Julian Wilson, European Commission – Joint Research Center
Kristina Wyatt, Persefoni
Leehi Yona, Stanford University
Daniel Zavala-Araiza, Environmental Defense Fund

Appendix D

Biographical Sketches of Committee Members

Donald J. Wuebbles (*Chair*) is Emeritus Professor of Atmospheric Science at the University of Illinois. He is also director of climate science for Earth Knowledge. From 2015 to 2017, Dr. Wuebbles was Assistant Director with the Office of Science and Technology Policy at the Executive Office of the President. After many years at Lawrence Livermore National Laboratory, Dr. Wuebbles came to the University of Illinois as professor and head of the Department of Atmospheric Sciences in 1994. He also led the development of the School of Earth, Society, and Environment at the University, and was its first director. Dr. Wuebbles is an expert in atmospheric physics and chemistry, with over 500 scientific publications related to the Earth's climate and atmospheric composition. He also provides analyses and development of metrics for translating science to policy and societal responses. He has been a leader in many international and national scientific assessments, including as a coordinating lead author on international climate assessments led by the Intergovernmental Panel on Climate Change (IPCC), thus contributing to IPCC being awarded the Nobel Peace Prize in 2007. He co-led Volume 1 of the 2017 4th U.S. National Climate Assessment. Among his major awards, Dr. Wuebbles has received the Cleveland Abbe Award from the American Meteorological Society, the Stratospheric Ozone Protection Award from the U.S. Environmental Protection Agency, and the Bert Bolin Global Environmental Change Award from the American Geophysical Union. He is a fellow of three major professional science societies: the American Association for the Advancement of Science, the American Geophysical Union, and the American Meteorological Society. Dr. Wuebbles holds a B.S. and M.S. in electrical engineering from the University of Illinois and a Ph.D. in atmospheric sciences from the University of California, Davis. He was a member of the joint U.S. National Academy of Science and UK Royal Society Committee on Climate Change that wrote *Climate Change: Evidence and Causes* in 2014 and updated in 2020.

Kamal Bawa (**NAS**) is Distinguished Professor Emeritus at the University of Massachusetts Boston, and founder-president of the Ashoka Trust for Research in Ecology and the Environment, Bengaluru, India. His main area of expertise is sustainability science. Among the many awards he has received are the Pew Award in Conservation and the Environment, Giorgio Ruffolo Fellowship at Harvard University, the Gunnerus Prize in Sustainability Science from the Royal Norwegian

Society of Letters and Sciences, the international MIDORI Prize in Biodiversity from the Aeon Foundation in Japan, the Linnean Medal from the Linnean Society, and honorary doctorates from the University of Alberta and Concordia University in Montreal. He is an elected Fellow of the U.S. National Academy of Sciences, the Royal Society, the American Philosophical Society, and the American Academy of Arts and Sciences. Dr. Bawa is the founding editor-in-chief of *Conservation and Society* and editor of *Ecology, Economy and Society*. Dr. Bawa received his Ph.D. in botany from Punjab University, India. He recently served on the National Academies' Committee on Developing a Booklet on Biodiversity for the Public and Policy Makers.

Gabrielle Dreyfus is chief scientist at the Institute for Governance & Sustainable Development (IGSD), Washington, DC, and Paris, and an adjunct faculty at Georgetown University. She joined IGSD in 2017 after nearly a decade of working at the science and policy interface with the U.S. Department of Energy, rising to deputy director for the Office of International Climate and Clean Energy, and previously with the National Oceanic and Atmospheric Administration and the U.S. Senate. In addition to dozens of scientific and technical publications, Dr. Dreyfus worked as the lead coordinating author on a synthesis report by the International Energy Agency and United Nations Environment Programme on the intersection of energy efficiency and the phasedown of hydrofluorocarbons in the cooling sector. Dr. Dreyfus is a member of the Montreal Protocol's Technology and Economic Assessment Panel Foams Technical Options Committee and Energy Efficiency Task Force and was a member of the technical review committee of the Global Cooling Prize. She also contributed to the design and implementation of the $50 million Kigali Cooling Efficiency Program (now Clean Cooling Collaborative), a philanthropic collaboration housed at the ClimateWorks Foundation. She was a 2021 Honoree of Environment+Energy Leader 100. Dr. Dreyfus holds a B.A. in Earth and planetary sciences from Harvard University, and a master's and Ph.D. from Princeton University and Sorbonne Université in geosciences.

Annmarie Eldering has over 30 years of experience in the fields of air pollution, greenhouse gases, and remote sensing. She retired from NASA's Jet Propulsion Laboratory (JPL) in 2022. Her early work at Caltech focused on measuring and modeling the aerosols that form in the Los Angeles Basin and drastically reduce visibility. Dr. Eldering developed advanced computer models to simulate these processes and evaluate possible strategies for emissions reductions and improvement in air quality. At the JPL, she was the Project Scientist on satellite projects to measure tropospheric air pollution (TES) and later carbon dioxide (the Orbiting Carbon Observatories (-2 and -3)). Through these satellite projects, she has worked closely with the modeling community that is combining ground-based measurements, emissions inventories, and satellite measurements in atmospheric models to create the most complete understanding of the carbon cycle and the state of Earth's atmospheric composition including greenhouse gases. She was also a deputy section manager and section manager in Earth Atmospheric Science at JPL for 5 years, guiding and organizing a cohort of near 100 scientists, technical staff, and postdocs. Dr. Eldering received her B.E. in chemical engineering from Cooper Union and her Ph.D. in environmental engineering science from Caltech.

Fiji George has more than 27 years of experience covering corporate climate and sustainability strategy, fundamental science, regulatory and policy experience along natural gas value chain-exploration/production, gas processing, transmission, and storage, and liquefied natural gas (LNG). His expertise focuses on researching and implementing sustainable solutions for prudent development and use of natural gas and LNG in a low-carbon economy, and integrating corporate environmental, social, and governance (ESG) programs to support the energy transition. Mr. George was a member of the National Academies of Sciences, Engineering, and Medicine Committee on Anthropogenic Methane Emissions in the United States: Improving Measurement, Monitoring,

Reporting, and Development of Inventories. He is a co-author on multiple peer-reviewed scientific papers, and the architect of the ONE Future Coalition voluntary methane inventory and mitigation program design. He has participated at Intergovernmental Panel on Climate Change (IPCC) Expert Meetings for Technical Assessment of IPCC Inventory Guidelines and provided feedback to the U.S. Environmental Protection Agency on annual U.S. national greenhouse gas inventories. At Cheniere, he leads the development of corporate policies and positions on climate and sustainability issues, including integration of climate considerations into corporate strategies and novel energy transition business plans such as the Cargo Emissions Tag, Quantification, Monitoring, Reporting, and Verification, and Lifecycle Analysis.

Heather Graven is a reader in the Department of Physics and the Grantham Institute at Imperial College London, United Kingdom. She earned her Ph.D. in Earth sciences from Scripps Institution of Oceanography at the University of California, San Diego, and she worked previously at ETH Zurich, Switzerland. Her research focuses on the use of atmospheric measurements to understand the global carbon cycle and its response to human activities and climate change. She uses radiocarbon and stable carbon isotopes to distinguish fossil fuel and biogenic influences on carbon dioxide and methane and to investigate carbon cycling in the ocean and land biosphere.

Kevin Gurney is an atmospheric scientist, ecologist and policy scientist currently working in the areas of carbon cycle science, climate science, and climate science policy at Northern Arizona University (NAU) where he is a professor in the School of Informatics, Computing, and Cyber Systems. Dr. Gurney is a cofounder of a for-profit business, Crosswalk Labs, which licenses a product that generates estimates of CO_2 emissions at fine scales across the United States to provide high-resolution CO_2 emissions to businesses, media, states, and local jurisdictions. Dr. Gurney's current university research involves understanding elements of the global carbon cycle using a variety of data/model fusion approaches. Over the last two decades Dr. Gurney has focused on quantification of fossil fuel CO_2 emissions at the global (FFDAS), national (Vulcan), and urban (Hestia) scales. Using data mining and assimilation algorithms, these very high-resolution greenhouse gas quantification products are being used by analysts, scientists, and governments for emissions mitigation planning, tracking, and assessment. The U.S. work, in particular, is anchoring efforts at the National Oceanic and Atmospheric Administration and the National Institute of Standards and Technology to develop prototypes of a multiscale greenhouse gas information system. He has degrees from the University of California, Berkeley, the Massachusetts Institute of Technology (MIT), and Colorado State University. He previously held faculty positions at Purdue University and Arizona State University. Dr. Gurney is an Intergovernmental Panel on Climate Change lead author, a National Science Foundation CAREER award recipient, Sigma Xi Young Scientist recipient, a Fulbright scholar, NAU Research and Creativity awardee, and has published over 150 peer-reviewed scientific articles with multiple papers in journals such as *Nature* and *Science* and a book from MIT Press, *Mending the Ozone Hole*.

Angel Hsu is an assistant professor of public policy and the environment at the University of North Carolina, Chapel Hill, and founder/director of the Data-Driven EnviroPolicy Lab, an interdisciplinary research group that innovates and applies quantitative approaches to pressing environmental issues. Her research explores the intersection of science and policy and the use of data-driven approaches to understand environmental sustainability, particularly in the areas of climate change and energy, urbanization, and air quality. Dr. Hsu has provided expert testimony to the U.S. Senate Committee on Energy and Natural Resources, U.S.-China Economic Security and Review Commission and is a member of the National Committee on U.S.-China Relations and a Public Intellectual Program Fellow. She is a lead and contributing author to global climate science assessments, including the IPCC Sixth Assessment Report and the United Nations Environment Programme (UNEP) Emissions Gap Report. She previously held a joint appointment as assistant professor of environmental studies at Yale-NUS College in Singapore and the Yale School of Forestry and Environmental Studies as an adjunct. She holds a Ph.D. in environmental policy from Yale University, an M.Phil. in environmental policy from the University of Cambridge, and a B.S. in biology and B.A. in political science from Wake Forest University in Winston-Salem, NC.

Tomohiro Oda is a senior scientist at the Universities Space Research Association and an adjunct professor in the Department of Atmospheric and Oceanic Science at the University of Maryland, College Park. Prior to his current positions, Dr. Oda held positions at NASA Goddard Space Flight Center, Colorado State University, NOAA Earth System Research Laboratory, and Japan's National Institute for Environmental Studies. Dr. Oda is a pioneer of the use of Earth observations in development of high-resolution carbon spatial emission estimates. During his career, he has advanced spatial carbon emission modeling and the associated error and uncertainty quantification. Dr. Oda has also advanced and matured the carbon emission quantification using new space-based carbon observation. Dr. Oda is the Principal Investigator and developer of the Open-source Data Inventory for Anthropogenic CO_2 (ODIAC) emission inventory data product. He is a science team member of NASA's Orbiting Carbon Observatory mission as well as a key contributor to Japan's Greenhouse gas Observing Satellite (GOSAT) mission. Dr. Oda holds a Ph.D. and a master's degree in engineering from Osaka University, Japan, and an undergraduate degree in physics from Kwansei Gakuin University, Japan.

Irène Xueref-Remy is Physicist of the National Council of Astronomers and Physicists, France, and Principal Investigator of the atmospheric activities at the Center for Scientific Research Observatoire de Haute Provence. She is a professor at the University of Aix-Marseille and she directs research programs at the Mediterranean Institute of Biodiversity and marine and continental Ecology at Aix-en-Provence. Her research focuses on assessing the variability of atmospheric greenhouse gases at different spatiotemporal scales; improving knowledge of emission inventories and natural greenhouse gas fluxes through atmospheric measurements and modeling techniques; and bridging the gap between science and society by communicating her knowledge through conferences and collaborating with local air quality agencies and stakeholders within her research projects. Her current research focuses on the construction of scenarios for reaching carbon neutrality by 2050 in the Aix-Marseille metropolis in collaboration with local stakeholders. Among others, she is a member of the Scientific Council of the Regional Air Quality Agency and a member of the WMO IG^{3}IS international program (Integrated Global Greenhouse Gas Information System) dedicated to help guiding stakeholders for taking valuable GHG emission-reduction actions in response to climate change. Dr. Xueref-Remy has a diploma of Engineer from the Ecole Centrale de Lyon, obtained her Ph.D. at the University of Grenoble, and performed three postdoctoral positions at the Juelich Research Center in Germany, Harvard University in the United States, and Laboratoire des Sciences du Climat et de l'Environnement in France.

Appendix E

Disclosure of Unavoidable Conflicts of Interest

The conflict-of-interest policy of the National Academies of Sciences, Engineering, and Medicine (https://www.nationalacademies.org/about/institutional-policies-and-procedures/conflict-of-interest-policies-and-procedures) prohibits the appointment of an individual to a committee like the one that authored this Consensus Study Report if the individual has a conflict of interest that is relevant to the task to be performed. An exception to this prohibition is permitted only if the National Academies determine that the conflict is unavoidable and the conflict is promptly and publicly disclosed.

When the committee that authored this report was established a determination of whether there was a conflict of interest was made for each committee member given the individual's circumstances and the task being undertaken by the committee. A determination that an individual has a conflict of interest is not an assessment of that individual's actual behavior or character or ability to act objectively despite the conflicting interest.

Mr. George was determined to have a conflict of interest because of his current employment at Cheniere Energy Inc. and his ownership of shares in Kinder Morgan.

Dr. Gurney was determined to have a conflict of interest because he is a co-founder of Crosswalk Labs, which distributes a data product of estimated CO_2 emissions.

In each case, the National Academies determined that the experience and expertise of the individuals were needed for the committee to accomplish the task for which it was established. The National Academies could not find another available individual with the equivalent experience and expertise who did not have a conflict of interest. Therefore, the National Academies concluded that the conflict was unavoidable and publicly disclosed it on its website (www.nationalacademies.org).